# How
# TO CHOOSE
# BINOCULARS

by ALAN R. HALE

Published by C & A Publishing
800 S. Pacific Coast Hwy., Suite 8-111
P.O. Box 8015-111
Redondo Beach, CA 90277

ISBN 0-9629090-0-9

# ACKNOWLEDGEMENTS

To Cherie, my wife, who had to put up with
long nights of work by me in writing this book.

To my children;
Joseph, Michael, Lara,
Lisa and David.

To Barbara Ann Colón for her patience
in inputting data in the computer with
constant changes.

To Linda Waible for editing.

To Ashby Spratley for the line drawings.

To all the binocular suppliers who assisted me
with information and photographs.

# How
## TO CHOOSE
## BINOCULARS

# PREFACE

Why write a book on binoculars? You might think that a binocular is something that you pick up, aim and look through. Well, there is more to it than that.

Being in the optical business for almost thirty years, I have gained an appreciation for good optical instruments. Binoculars are no exception and all binoculars are *not* created equally — there are excellent ones, good ones and very poor ones.

Depending on what you will use the binoculars for, there are certain specifications you need to know. Unfortunately, many manufacturers do not tell you the specifications (some don't want you to know) that may be important but they do give you a lot of marketing hype and pretty pictures.

You might think that you can go to the library or bookstore and learn about binoculars, as I tried. However, it's amazing that there are virtually no books on the subject. True, there are many magazine articles periodically on binoculars but the subject matter is short, many times with erroneous or misleading information.

Therefore, in an effort to share with you the basics of binoculars, who offers them and what the specifications are that you should know, this book was written.

# How
## to choose
## binoculars

# CONTENTS

# CHAPTER 1

## HOW TO CHOOSE BINOCULARS

It is very easy to choose a pair of binoculars for many people — just select the lowest price, a fancy model name along with an attractive color and style. What more is important? A lot! What you are going to do and how you are going to use the binoculars will help determine the best binocular for you.

Binoculars appear to be simple optical devices but in reality, they are complex, precision optical instruments. An educated consumer will be much happier learning about the various aspects of binoculars *before* making a purchase. This book tells you the basics of binoculars and gives you a lot of information to digest before making a decision.

In general, with binoculars you usually get what you pay for. As the price increases, in most cases, so does the quality of the unit. In my opinion, most binoculars with a list price of under $100.00 (there are a few exceptions) are not worth considering. Beware of units advertised at $19.95 or $29.95 — this is indeed too good to be true!

There is a multitude of binoculars available in the marketplace from dozens of manufacturers and suppliers. It can be very confusing trying to sort through the maze. The models I have included in the various tables are all from reputable firms. Be careful of unknown names.

I cannot tell you which particular binocular is best for your application as *only you can choose* the unit best suited for your particular purpose and usage. This book can be a guide to help you make your choice. It is amazing that in most binocular brochures you are told that model "X" is the best for birding, model "Y" is best for sporting events, model "Z" is best for hunting, etc. This is ridiculous because it depends on exactly what you are going to be doing and under what conditions.

It also doesn't help you that most store clerks know virtually nothing about the binoculars they are selling. So, read on and gain some knowledge about binoculars.

# CHAPTER 2

## WHAT IS A BINOCULAR?

Simply, a binocular is two low powered telescopes mounted together. As compared to a telescope, binoculars allow you a more comfortable viewing experience by using both eyes in a relaxed manner with a three-dimensional image.

Any binocular has two basic functions:

1 To gather more light than the unaided human eye does.
2 To enlarge the image of what you're looking at.

However, a binocular is much more than a basic definition.
A binocular can:

- open up new vistas of learning
- add pleasure to your life
- create excitement with new discoveries
- add new dimensions to whatever you enjoy looking at
- expand hobbies to gain more fulfillment
- aid in your appreciation of the world
- allow families or friends to share the
  enjoyment of observing

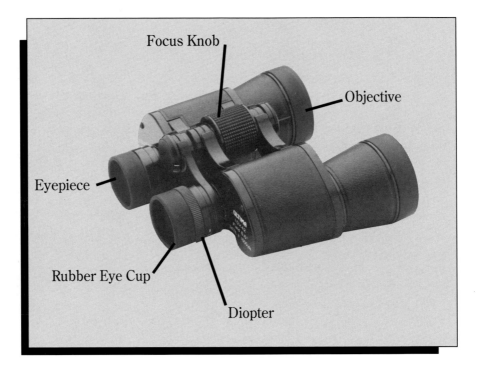

Focus Knob

Objective

Eyepiece

Rubber Eye Cup

Diopter

# CHAPTER 3

A binocular can be used to look at virtually anything that you look at with your eyes and can greatly enhance that view. Possible applications include:

- Astronomy
- Backpacking
- Birding
- Boating
- Camping
- Concerts
- Fishing
- Flying
- General Purpose
- Hiking
- Hunting

- Long Distance Observations
- Museum Visits
- Nature Study
- Sporting Events
- Surveillance
- Target Shooting
- Theater/Opera
- Travel
- View Home Applications
- Voyeurism

# CHAPTER 4

As I mentioned earlier, it depends on the circumstances. I will give you a few examples of the wide range of uses:

For birding, an 8x21 model may be just fine on a bright day for observing a hummingbird at your bird feeder 10 to 20 feet away. But if you are observing an eagle's nest 1/2 mile away, this may call for a 20x80 model on a tripod. And if you are out on a field trip in a dark forest, you may need a pair of 10x42 or 8x56 binoculars.

If you are in a large stadium watching a football game, it depends on your seat location (are you at the 50 yard line in the first few rows or are you behind the end zone at the top of the stadium?). 7x35's might be your choice for 50 yard line viewing, 10x50's for the end zone. It also depends on whether the game is mid-afternoon or a night game. The night game is better seen with larger objectives.

For astronomy, a pair of 7x35 binoculars may be fine for observing the Moon or wide star fields. However, for detailed observations of the Moon or to see more details on planets, you need higher power and larger objective lenses.

If you are backpacking, you may want to keep your load light and carry a pair of compact binoculars (8x21, 8x25 or even a monocular) but if you are serious about observing distant scenery or wildlife, you may not mind the extra weight of a 7x42 or 7x50 binocular.

For the mariner, most manufacturers say that 7x50 is the ideal unit. This is a good low light performing binocular but if great distances are what you're after, then a 10x50 might be better. You also must consider the weight in which case an 8x21 or 8x25 might be best for you.

So, you can see from these brief examples that the size binocular you choose depends on the circumstances. In many cases, you will find that a single pair of binoculars is not enough and you may eventually want two, three or more different binoculars for various applications.

# CHAPTER 5

The majority of binoculars of medium to high quality sold in the United States are made in Japan with a smaller percentage from South Korea. High priced and high quality binoculars are also available from Germany and Austria. A lot of binoculars come from Taiwan, Hong Kong and other Asian countries but at this time they are generally of poor quality and resultant low prices.

A few million binoculars are sold each year and have been for many years, making it a fairly large industry.

Numerous manufacturers and suppliers are engaged in providing binoculars. This book lists 26 manufacturers (or their agents) which represent established, fairly well-known brands of high reputation. Probably well over 1000 models (ranging from 5x20 to 40x150) are available in the U.S.A. I have selected approximately 400 models, the majority of these having suggested list prices over $100.00. I chose only companies that had at least six models in their line and did not include any private label models from large mass merchandisers.

# CHAPTER 6

Being in the binocular business, I have always kept a file of suppliers' catalogs, brochures and advertisements. During industry trade shows I examine the various binoculars offered.

What is available to the consumer in regard to specifications is limited from most suppliers. In order to provide as much information as possible to help consumers in the buying process, I contacted the top people at each company with my requests for additional information. My thanks to all those who complied.

Due to the vast amount of data compiled, I cannot guarantee 100% accuracy of all data.

# CHAPTER 7

## CRITERIA FOR CHOOSING BINOCULARS

In deciding on a pair of binoculars there are many factors to consider, including;

- power
- objective lens size
- field of view
- exit pupil
- eye relief

- prisms
- type of coatings
- near focus
- weight
- appearance

- and more

Many of these factors are related and affect the others to some degree.

This chapter will describe the various binocular criteria to help you in making your choice or choices.

## METRIC CONVERSION

Since binoculars are supplied from manufacturers outside the U.S.A., most specifications are in metric form; the following table may be helpful as you read this book if you are not familiar with the metric system:

| Metric | Approx. U.S. Equivalent |
|--------|-------------------------|
| millimeter (mm) | 0.039 inches |
| meter (m) | 39.37 inches |
| gram (g) | 0.035 ounces |

For example, 42mm = 1.638 inches (42x0.039); and 100 meters = 3937 inches (100x39.37) or 328.08 feet or 109.36 yards; and 325 grams = 11.38 ounces (325x0.035).

## POWER (MAGNIFICATION)

The power (or magnification) of a binocular means the degree to which the object or subject you are looking at is enlarged. For example, with a 7x42 binocular, the first number (7) is the binocular power. Virtually all binoculars sold have the power indicated on the binocular. A binocular of 7 power enlarges (or magnifies) the image seven times the size as seen by the normal, unaided human eye. You can also look at it as if the image were only 1/7th of the actual distance away. If the object you were looking at was 700 feet away, with a 7x binocular, it would appear as though it was only 100 feet away.

More power does not mean better. How much power you need depends on several factors. Binoculars over 10 power usually require a photographic/video tripod mounting for stability as they are too hard to hand-hold. However, in the last decade we find offered many 10 and even 12 power binoculars which can be handheld due to weight reductions. In general, high powers are limited to specialized areas of interest such as astronomy, long distance observing or for surveillance work.

Keep in mind that power affects brightness and the lower the power, the brighter the image (although objective size, exit pupil, type of coatings and type of prisms also affect brightness).

In general, increasing power will reduce field of view and eye relief.

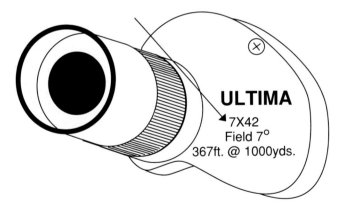

Drawing showing (typically) where the power is indicated on a binocular

**1x** (Unaided Eye)

Photos showing how power changes the object size

The objective lenses of binoculars are the front lenses and their diameters are specified in millimeters. For example, with a 7x42 binocular, the second number (42) is the diameter of the objective lenses. Virtually all binoculars sold have the objective size indicated on the binocular.

The purpose of the objectives is to gather incoming light and form a sharply focused image of distant subject matter.

The larger the objective's diameter, the more light the binoculars gather and the brighter resultant image with more detail and sharpness. This effect is most pronounced under low light conditions (dawn, dusk, overcast day, thick forest) or nighttime use.

If you double the size of the objectives, you quadruple the light gathering ability. For example, a 7x50 binocular has almost twice the light gathering ability of a 7x35 binocular and has four times the light gathering ability of a 7x25 binocular.

Most high quality binoculars utilize two-element, achromatic objective lenses. The two elements are of different types of glass to eliminate color errors and most binoculars (other than very inexpensive ones) do a good job of this.

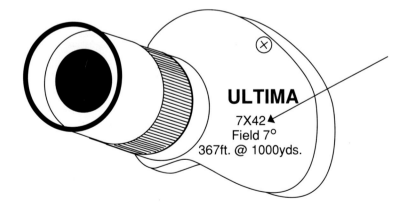

ULTIMA
7X42
Field 7°
367ft. @ 1000yds.

Drawing showing (typically) where the objective size
is indicated on a binocular

Erecting prisms are used in binoculars to correct an inverted image.

There are two types of prism systems: (1) the Porro prism; and (2) the roof (or Dach) prism. Which is best is a matter of disagreement. In general, Porro prisms are thought to yield superior optical performance but the roof prisms utilized by Zeiss, Leica and a few others provide excellent performance due to their phase shift coating (see page 37). Whichever type is used can perform well assuming that they are well-made and care is taken in the manufacturing process.

Porro prisms come in basically two common styles: (1) BK-7 and (2) BAK-4. The main difference between the two styles is the glass density (or refractive index). BK-7 utilizes boro-silicate glass and BAK-4 uses barium crown glass. The BAK-4 is the finer glass (a higher density) and eliminates internal light scattering and produces sharper images than the BK-7 glass. BAK-4 prism usage results in higher prices for binoculars than ones using BK-7 prisms.

A few other glass types are used for prisms but unless the density is higher than BAK-4, they are not as good as density is the most important factor.

Many manufacturers do not tell you the type of Porro prisms being used. The use of BK-7 glass prisms can usually be determined by holding the binoculars away from your eyes at a bright light source to see the cone of light coming through. With BK-7 you will note a square shape cutting off part of the cone (except on very narrow field binoculars) whereas with the BAK-4 you generally see the complete round cone of light. Thus, the BK-7 loses some of the light that strikes near the edges of the prisms. On some units using BAK-4 prisms, a slight cut-off of the edges (or one edge) is apparent but no noticeable light loss occurs.

Roof prisms by design are lighter in weight and more compact. They are more complex and difficult to manufacture with more precise tolerances than Porro prisms and thus generally cost more. For example, the angle of the roof apex must be exact at 90° or this can cause severe problems.

Roof prisms that are undersize do not produce complete circular round exit pupils and some light is lost. This effect can also be caused by mechanical obstructions in the light path.

## Roof Prism

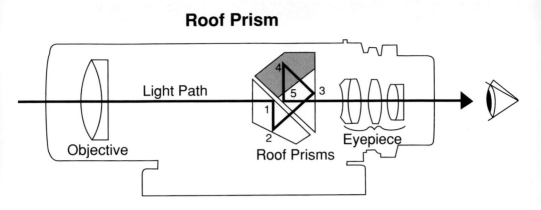

## Porro Prism

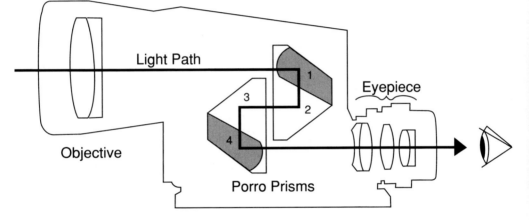

## Galilean

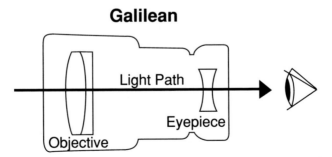

Drawing showing prism designs (Porro and Roof)
as well as non-prism type (Galilean)

## PHASE SHIFT COATING OF ROOF PRISMS

In normal roof prisms, due to the nature of light waves, 70% of the light reflected off of one roof surface is 1/2 a wavelength phase shifted from the reflected light off of the other roof surface. This causes a deterioration in contrast. Because of this problem, Porro prism binoculars have had an edge in optical quality.

However, during the last few years, some manufacturers have begun coating the roof with an anti-phase shifting material. Thus, a phase corrected roof prism binocular is equal to (not better or worse) a Porro prism binocular in optical performance assuming all else is equal.

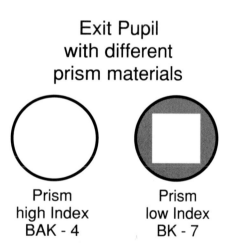

### Exit Pupil
### with different
### prism materials

Prism
high Index
BAK - 4

Prism
low Index
BK - 7

Drawing showing the field using BAK-4 and BK-7 prisms

## EYEPIECES

The eyepieces (oculars) are the lenses closest to your eyes. The eyepieces enlarge (or magnify) the image formed from the objectives after it has passed through the prisms of the binocular. They generally consist of three to six optical elements or more.

The Kellner eyepiece is the one most often used in binoculars. To obtain a wider field of view, the Erfle eyepiece is usually used. An Erfle has a more complex design and a resultant higher price. Koenig eyepieces (high quality) are also utilized in some units.

The specific type of eyepiece used is not important. Each design has had many different modifications and thus many hybrid designs exist. **The most important factor in eyepieces is the quality and care of manufacturing and the type of optical coatings applied.**

## OPTICAL COATINGS

Coatings of the optical elements of binoculars (objective lenses, oculars and prisms) reduce light loss and glare due to reflection and increase light transmission and contrast.

Reflected light is a limitation of binoculars. When light strikes normal glass, 4 to 5% of it is reflected back from the glass surface (glass itself absorbs some light as it passes through). With from 10 to 16 glass surfaces in an average pair of binoculars, it is possible to lose 50% or more of the light originally striking the objective lens. Worse yet, there is all that reflected light bouncing around inside your binoculars which causes glare and ghost images.

If the surface of the glass is coated with a very thin vacuum film of certain chemicals (usually magnesium fluoride - $MgF_2$), the reflection of light can be reduced to about $1\frac{1}{2}$ to 2% per glass surface. This is called an antireflection coating.

Multi-coatings (multi-films of chemicals) can reduce the amount of reflected light to 0.25 to 0.5% per glass surface and result in a higher light transmission and better contrast than standard coatings.

Coatings must be absolutely uniform in thickness and density or unpredictable reflections and other problems occur.

Many people think they can tell how good coatings are by their color. This is not possible and the color of the coatings can vary enormously by the procedures used. Magnesium fluoride color ranges from a pale blue to a deep purple violet. Multi-coatings can exhibit various colors or multi-. colors depending on the angle you are looking at and can be blue, purple, violet, green, red or yellow.

Occasionally you will see a pair of binoculars designated as having UV coatings. These reduce glare and internal light reflections. They do reject harmful UV light and perform well at high altitudes but alone do not transmit as much light as multi-coatings.

**Coatings are one of the most hyped, and in many cases, most misleading specifications of binoculars. The various terms below are general descriptions for comparison sake:**

"coated optics" (C) — means only that one or more surfaces of one or more lenses has received an anti-reflective coating.

"fully coated" (FC) — means that all air-to-glass surfaces have been coated (but in many cases it means something less). If any plastic lenses are used they are most likely not coated.

"multi-coated" (MC) — means that one or more surfaces of one or more lenses have been coated with multiple films. Some surfaces could be single coated or some not coated at all.

"fully multi-coated" (FMC) — means that all air-to-glass surfaces should have received multiple films.

Roof prism binoculars usually have one surface of each prism aluminized (or silvered) which theoretically means that they have more light loss than Porro prism type binoculars but it is insignificant.

If you can afford it, buy binoculars that are fully multi-coated as the light transmission will be higher and the image brighter. Most reputable companies offer high quality multi-coatings with various fancy marketing names but the results will be worth it. It doesn't matter if one company offers a couple of percentage points higher transmission than another because you will not be able to detect any difference in usage.

## AN EXAMPLE OF LIGHT LOSS

A typical binocular with 14 glass surfaces has the following total light loss on average:

| No Coatings | FC | MC | FMC |
|:---:|:---:|:---:|:---:|
| 47% | 17% | 11% | 5% |

From the above, if this binocular is FMC, the throughput light transmission would be about 95%. This is excellent and about the best that can be done. I would be skeptical of firms who advertise higher rates.

## FIELD OF VIEW

The field of view is the size, in degrees (called the angular field), of the area you can see with a pair of binoculars. The image you see is circular. Sometimes the field of view is expressed as the width, measured in feet, of the viewing area you would see at 1000 yards (linear field). Most binoculars sold have the angular and/or linear field indicated on the binocular.

If you only know the angular field but want to know the linear field, then multiply the angular field by 52.5. For example, if the angular field is 8°, then 8x52.5=420 ft. (linear field). In reality 1°=52.365 ft. @ 1000 yards but in the industry it is rounded to 52.5.

**Linear Field = Angular Field x 52.5**

The greater the field of view, the greater the area you will see in the image.

Greater fields of view are not always better but helpful for spectator sports, nature watching, hiking and astronomy. Greater field of view is important where the object being observed is likely to move or when you are moving. However, in general, wider fields usually mean less eye relief and more money due to the complexity of wide field eyepieces.

Field of view is related to power such that the greater the power (in general), the smaller the field of view. However, the type of eyepieces used and the focal length of the objectives determine the actual field of view. The diameter of the objectives has no bearing on the field of view.

Some binoculars are designated as wide angle models. There is no real criteria to determine what is called wide angle but generally it is accepted that any binocular with eyepieces having 65° or more of apparent field is labeled as wide angle. This is the Japanese standard and the Europeans consider 60° as being wide angle.

The apparent field is defined as the angle your eye covers from one side to the other of the subject matter in the eyepiece. The apparent field can be determined by the formula: apparent field = field of view (in degrees) times power. For example, a 7x42 binocular with a 7° field of view has an apparent field of 49° (7x7).

**Apparent Field  =  Angular Field  x  Power**

The field of view of the naked eye is 180°. But in reality we can only see about 140°.

The full moon is 1/2° across (the same area as a small human finger held at arm's length).

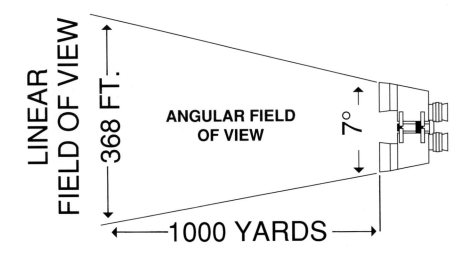

**LINEAR FIELD OF VIEW**

368 FT.

**ANGULAR FIELD OF VIEW**

7°

←——1000 YARDS——→

Drawing showing angular and linear fields of view of a typical binocular
(7 x 42 with a 7° field of view)

Photograph showing what differences in field of view mean

The exit pupil is the size (or diameter) of the beam of light (in millimeters) that leaves the eyepiece of the binocular. The larger the exit pupil, the brighter the image.

Large exit pupils are advantageous when viewing in low light conditions or at nighttime.

For astronomical applications, the rule has been that the exit pupil should correspond with the dilation of the dark-adapted pupil of your eye, which is between 5 and 9mm (maximum available). This maximum size decreases with age. Also the size is diminished by smoking, contaminants in the air, poor health and possibly by bright starlight. This same rule also applies to other binocular uses in low light conditions.

Different experiments by Lowenfeld and Kornzweig have shown how our pupils decrease with age (see the accompanying graphs). People of all ages vary tremendously in their pupil size but for most people, the pupil size decreases from the mid-teens on. As we grow older, the dark-adapted pupil is reduced and grows closer to the same size it is in bright light. However, there are many exceptions to the experiments.

On a bright day, your pupil shrinks to about 2 (high noon sun) to 4mm. Then you would see no difference in brightness between a 2mm exit pupil and a 9mm exit pupil.

If the exit pupil of a binocular is larger than the entrance pupil of the eye, some light will not be able to enter the eye. The image seen will continue to have the same brightness regardless of how much the size of the exit pupil exceeds the entrance pupil of the eye. The exception is in astronomy when looking at point sources (stars) where the brightness of the star's image is determined by the size of the objective.

Even in bright light a 7mm or larger exit pupil can be helpful if you are moving or on a boat by making it easier to keep your eyes' daylight-contracted pupil centered in the binocular's larger exit pupil.

You can see the exit pupil by holding your binoculars out at arm's length and looking through the eyepieces at a light source.

To calculate the exit pupil, divide the objective by the power. For example, the exit pupil of 7x42 binoculars is 42 divided by 7 = 6mm.

$$\text{Exit Pupil} = \frac{\text{Objective Diameter (mm)}}{\text{Power}}$$

The accompanying table shows the exit pupil size of the various binoculars listed in this book. They range from 1.4 to 7.3 mm.

Generally, exit pupil decreases with greater power (assuming the same size objective lenses). This is demonstrated with zoom binoculars. For example, with a 7-21x35 zoom binocular, the exit pupil at 7 power is 5mm and at 21 power it is decreased to 1.7mm.

| Size | Exit Pupil | Size | Exit Pupil | Size | Exit Pupil |
|------|-----------|------|-----------|------|-----------|
| 5x20 | 4.0 | 6x30 | 5.0 | 10x50 | 5.0 |
| 6x20 | 3.3 | 7x30 | 4.3 | 12x50 | 4.2 |
| 7x20 | 2.9 | 8x30 | 3.8 | 16x50 | 3.1 |
| 8x20 | 2.5 | 9x30 | 3.3 | 20x50 | 2.5 |
| 9x20 | 2.2 | 10x30 | 3.0 | 21x50 | 2.4 |
| 10x20 | 2.0 | 6x32 | 5.3 | 24x50 | 2.1 |
| 6x21 | 3.5 | 8x32 | 4.0 | 35x50 | 1.4 |
| 7x21 | 3.0 | 10x32 | 3.2 | 8x56 | 7.0 |
| 8x21 | 2.6 | 6x35 | 5.8 | 10x56 | 5.6 |
| 10x21 | 2.1 | 7x35 | 5.0 | 10x60 | 6.0 |
| 8x22 | 2.8 | 9x35 | 3.9 | 15x60 | 4.0 |
| 6x23 | 3.8 | 10x35 | 3.5 | 20x60 | 3.0 |
| 8x23 | 2.9 | 12x35 | 2.9 | 30x60 | 2.0 |
| 10x23 | 2.3 | 15x35 | 2.3 | 9x63 | 7.0 |
| 12x23 | 1.9 | 17x35 | 2.1 | 12x63 | 5.3 |
| 6x24 | 4.0 | 6x36 | 6.0 | 10x70 | 7.0 |
| 7x24 | 3.4 | 8x36 | 4.5 | 14x70 | 5.0 |
| 8x24 | 3.0 | 12x36 | 3.0 | 20x70 | 3.5 |
| 9x24 | 2.7 | 7x40 | 5.7 | 30x70 | 2.3 |
| 10x24 | 2.4 | 8x40 | 5.0 | 11x80 | 7.3 |
| 12x24 | 2.0 | 9x40 | 4.4 | 15x80 | 5.3 |
| 15x24 | 1.6 | 10x40 | 4.0 | 16x80 | 5.0 |
| 6x25 | 4.2 | 14x40 | 2.9 | 20x80 | 4.0 |
| 7x25 | 3.6 | 16x40 | 2.5 | 30x80 | 2.7 |
| 8x25 | 3.1 | 21x40 | 1.9 | 14x100 | 7.1 |
| 9x25 | 2.8 | 7x42 | 6.0 | 20x100 | 5.0 |
| 10x25 | 2.5 | 7.5x42 | 5.6 | 25x100 | 4.0 |
| 12x25 | 2.1 | 8x42 | 5.3 | 20x120 | 6.0 |
| 15x25 | 1.7 | 10x42 | 4.2 | 25x150 | 6.0 |
| 7x26 | 3.7 | 7.5x44 | 5.9 | 35x150 | 4.3 |
| 10x26 | 2.6 | 7x50 | 7.1 | 40x150 | 3.8 |
| 10x28 | 2.8 | 8x50 | 6.3 | | |

The exit pupil (in millimeters) is shown for the various sizes of binoculars including the low and high power of zoom binoculars

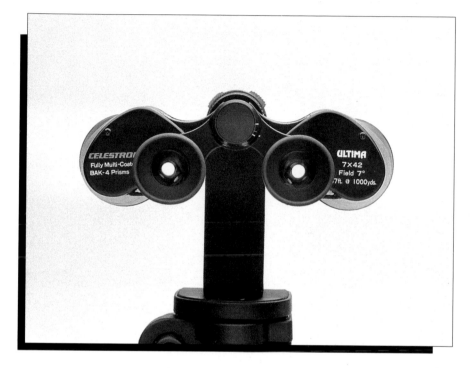

Photo showing the exit pupil on a 7x42 binocular (6mm)

# The Eye's Pupil Size

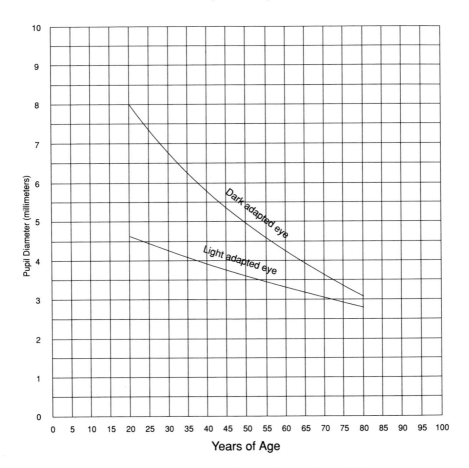

Adapted from <u>Night Vision</u> (National Academy Press, 1987)

Drawing showing the eye's pupil size

# The Eye's Dark - Adapted Pupil Size

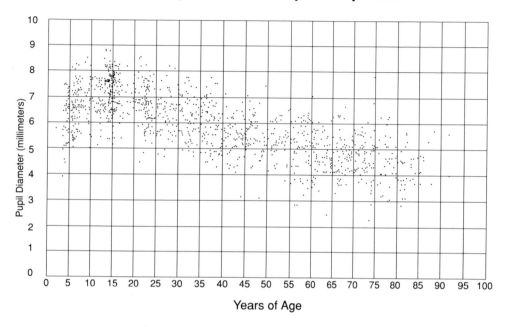

Pupil Diameter (millimeters) / Years of Age

Adapted from <u>Night Vision</u> (National Academy Press, 1987)

Drawing showing the dark-adapted pupil size

Eye relief is the distance a binocular can be held from the eye where you can observe the full field of view comfortably. This optimum position is called the eyepoint and the distance from the eyepiece lens is the eye relief measured in millimeters.

Longer eye relief provides viewing comfort and is especially helpful for eyeglass wearers. The distance from the average human eye to the inside of an eyeglass lens is about 12 to 14mm. It is about another 4 to 6mm to the plane of the binocular eyepiece lens. So for eyeglass wearers, you need a minimum of 16 to 20mm of eye relief to see the full field of view.

Some manufacturers use the term "high eyepoint" which means long eye relief.

Keep in mind that wide angle binoculars usually mean less eye relief.

Eyeglass wearers need longer eye relief but they also suffer from peripheral light loss. If you wear eyeglasses and do not have astigmatism, it is best (if possible) to remove your eyeglasses when using binoculars. If you must wear eyeglasses, then fold down the rubber eyecups (or other mechanism), if available, to obtain the widest field possible.

If you do not wear eyeglasses, longer eye relief is better since your eye comfort is increased as you do not have to press hard against the eyecups to see the whole field of view.

You can measure the eye relief with a simple test. Aim the binoculars at a brightly lit sky while moving a piece of waxed paper toward and away from the eyepiece. A spot of light will appear on the waxed paper. When this spot's diameter is smallest and the image sharpest and brightest, you've found the eyepoint. Then measure this distance from the wax paper to the eyepiece lens in millimeters.

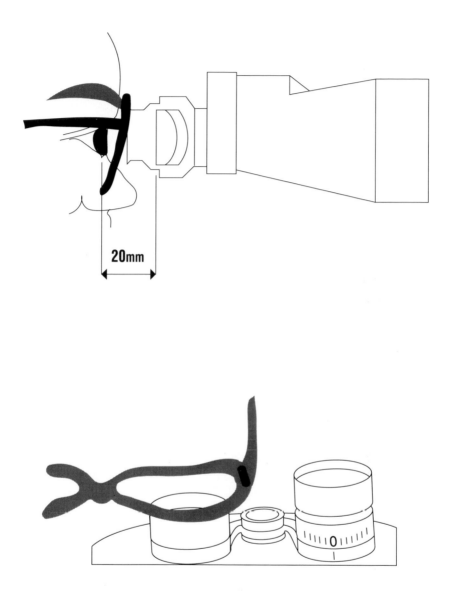

Drawing showing how wearing eyeglasses
changes your eye position when using binoculars

Near focus is the closest distance to the observed object that the binoculars can be used while retaining a sharp focus.

A binocular specification for near focus is a theoretical one based on a young person's eyes. Older people have a slightly more distant near focus. In addition, people in general have a wide variance in their particular eyes and near focus distances can vary from person to person. Also, eyeglass wearers who remove their glasses when using binoculars will note a change in near focus – near sighted people will note a closer near focus and far sighted people will note a further near focus.

Serious birders request a near focus of about 15 feet but 20-25 feet is usually acceptable.

To measure mathematically, find the adjustable range of the diopter and divide the square of the power by the positive figure of the measured range. For example, a particular 8x56 binocular has a diopter range of +5.5 to -5.5. Then $8^2$ = 64 divided by 5.5 = about 11 meters or about 36 feet for the near focus.

Some binocular repair companies can adjust particular binoculars to have a closer near focus but generally the infinity focus is compromised in optical sharpness.

## BRIGHTNESS

Brightness is a glasses' ability to gather and transmit enough of the available light to give a sufficiently bright image for good definition. Brightness also helps in differentiating colors of objects.

Brightness is dependent on several factors:

(a) Magnification applied to the image at the eyepiece

(b) Intensity of the light coming from the object you are viewing

(c) Size of the diameter of the objective as a light gathering lens

(d) The type and quality of glass used for the objective lens

(e) Transmission and reflection losses of light in passing
    through the instrument

# (1) LIGHT TRANSMISSION

Since the amount of light that falls on a circle is proportional to its area, and since the area is proportional to the square of the diameter, you can figure out how much light an objective lens will deliver to the eye by dividing the diameter of the objective lens (in mm) multiplied times itself, by the diameter of the pupil of the eye (in mm) times itself.

$$\frac{\text{objective diameter}^2}{\text{pupil diameter}^2}$$

For example with a pair of 8x40 binoculars and a pupil size of 5mm, we have the following:

$$8x40 \quad \text{or} \quad \frac{40x40}{5x5} = \frac{1600}{25} = 64$$

64 represents 64 times more light available than with the naked eye alone.

The formula above results in theoretical figures and does not take into account light loss from reflections.

Light transmission also can mean a percentage of how efficiently light is transmitted through the binocular's optical system. Unfortunately, there are no industry standards for light transmission. The best way of measuring is to use a scientific instrument called a photometer.

# (2) RELATIVE BRIGHTNESS INDEX (R.B.I.)

The R.B.I. is used as a comparison of image brightness but it does have severe limitations as discussed below. It is determined by squaring the exit pupil. An R.B.I. of 25 or more is considered useful in low light conditions. In bright light, a R.B.I. of 9 would have no brightness advantage over one of 50.

$$\text{R.B.I.} = \text{Exit Pupil (mm)}^2$$

To calculate the R.B.I. for a 7x42 binocular, take its exit pupil of 6mm (42 divided by 7). Then $6^2 = 36$ R.B.I.

One of the limitations of this index is that binoculars with different magnifications but identical exit pupil diameters have the same R.B.I. rating. We can use for comparison a pair of 7x35 and 10x50 binoculars. Both have an R.B.I. rating of 25 (both have exit pupils of 5). The 10x50's transmit more than twice the light to the image than the 7x35's and will provide a much brighter image despite their higher magnification.

Another limitation of the index is its assumption that all binoculars have the same light transmission which is not true.

So, the only usefulness to this index is in comparing R.B.I. for binoculars with the same size objective lenses. This being the case, you don't even need the formula as high magnifications yield lower brightness levels.

The accompanying table lists the R.B.I. for the various binoculars listed in this book.

## (3) TWILIGHT FACTOR

The twilight factor is a measurement of viewing efficiency (sharpness) and image detail in twilight (low light) conditions. The larger the twilight factor, the more efficient the low light performance. This test also has limitations.

To calculate the twilight factor, take the square root of the power times the objective lens diameter ($\sqrt{\text{power x objective}}$). For example, a 7x42 binocular has a twilight factor of 17.1. $\sqrt{7\text{x}42} = \sqrt{294} = 17.1$

**Twilight Factor** = $\sqrt{\textbf{Power x Objective Diameter (mm)}}$

This formula takes magnification more into account than the R.B.I. and is a better indicator of brightness. **It has been proven that when observing low contrast subject matter during twilight (not in the dark) higher magnifications will increase contrast and more detail will be seen.**

The problem with the twilight factor is the reverse of the R.B.I. Larger magnifications with larger objective lens diameters will end up with higher values. But no compensation is made for light loss with increasing magnifications nor does it take into account that all binoculars of the same size do not have equal light transmission. It is useful in comparing different objective lens diameters with the same magnification but is unnecessary since again we know that larger objective lens diameters have brighter images.

The accompanying table lists the twilight factor for the various binoculars listed in this book.

## (4)  RELATIVE LIGHT EFFICIENCY (R.L.E.)

The R.L.E. number is used to take into account the use of new prism materials, new objective lens materials and improvements in coatings, all of which affect the brightness. This formula has limitations similar to the R.B.I.

The formula arbitrarily fixes transmission of uncoated binoculars at 50%.

R.L.E. = R.B.I. + 50%. (to compensate for coated optics)
A binocular with an R.B.I. of 25 has a R.L.E. of 37.5.

BAK-4 prisms increase R.L.E. by 10 to 15% more and newer multi-coatings increase R.L.E. even more.

## ALL INDICES

All the various indices (R.B.I., Twilight Factor, R.L.E.) are guidelines to try and compare the brightness of different size binoculars. Their usefulness by themselves is questionable. More important are the objective lens diameter and secondarily, the magnification in finding the optimum amount of brightness and detail delivered to your eyes. Also to be considered are the type of prisms being used, as well as the type of optical coatings.

# R.B.I. OF VARIOUS BINOCULARS

| Size | RBI | Size | RBI | Size | RBI |
|------|-----|------|-----|------|-----|
| 5 x 20 | 16.0 | 6 x 30 | 25.0 | 10 x 50 | 25.0 |
| 6 x 20 | 11.1 | 7 x 30 | 18.4 | 12 x 50 | 17.4 |
| 7 x 20 | 8.2 | 8 x 30 | 14.1 | 16 x 50 | 9.8 |
| 8 x 20 | 6.3 | 9 x 30 | 11.1 | 20 x 50 | 6.3 |
| 9 x 20 | 4.9 | 10 x 30 | 9.0 | 21 x 50 | 5.7 |
| 10 x 20 | 4.0 | 6 x 32 | 28.4 | 24 x 50 | 4.3 |
| 6 x 21 | 12.3 | 8 x 32 | 16.0 | 35 x 50 | 2.0 |
| 7 x 21 | 9.0 | 10 x 32 | 10.2 | 8 x 56 | 49.0 |
| 8 x 21 | 6.9 | 6 x 35 | 34.0 | 10 x 56 | 31.4 |
| 10 x 21 | 4.4 | 7 x 35 | 25.0 | 10 x 60 | 36.0 |
| 8 x 22 | 7.6 | 9 x 35 | 15.1 | 15 x 60 | 16.0 |
| 6 x 23 | 14.7 | 10 x 35 | 12.3 | 20 x 60 | 9.0 |
| 8 x 23 | 8.3 | 12 x 35 | 8.5 | 30 x 60 | 4.0 |
| 10 x 23 | 5.3 | 15 x 35 | 5.4 | 9 x 63 | 49.0 |
| 12 x 23 | 3.7 | 17 x 35 | 4.2 | 12 x 63 | 27.6 |
| 6 x 24 | 16.0 | 6 x 36 | 36.0 | 10 x 70 | 49.0 |
| 7 x 24 | 11.8 | 8 x 36 | 20.3 | 14 x 70 | 25.0 |
| 8 x 24 | 9.0 | 12 x 36 | 9.0 | 20 x 70 | 12.3 |
| 9 x 24 | 7.1 | 7 x 40 | 32.7 | 30 x 70 | 5.4 |
| 10 x 24 | 5.8 | 8 x 40 | 25.0 | 11 x 80 | 52.9 |
| 12 x 24 | 4.0 | 9 x 40 | 19.8 | 15 x 80 | 28.4 |
| 15 x 24 | 2.6 | 10 x 40 | 16.0 | 16 x 80 | 25.0 |
| 6 x 25 | 17.4 | 14 x 40 | 8.2 | 20 x 80 | 16.0 |
| 7 x 25 | 12.8 | 16 x 40 | 6.3 | 30 x 80 | 7.1 |
| 8 x 25 | 9.8 | 21 x 40 | 3.6 | 14 x 100 | 51.0 |
| 9 x 25 | 7.7 | 7 x 42 | 36.0 | 20 x 100 | 25.0 |
| 10 x 25 | 6.3 | 7.5 x 42 | 31.4 | 25 x 100 | 16.0 |
| 12 x 25 | 4.3 | 8 x 42 | 27.6 | 20 x 120 | 36.0 |
| 15 x 25 | 2.8 | 10 x 42 | 17.6 | 25 x 150 | 36.0 |
| 7 x 26 | 13.8 | 7.5 x 44 | 34.4 | 35 x 150 | 18.4 |
| 10 x 26 | 6.8 | 7 x 50 | 51.0 | 40 x 150 | 14.1 |
| 10 x 28 | 7.8 | 8 x 50 | 39.1 | | |

The R.B.I. is shown for the various sizes of binoculars including the low and high power of zoom binoculars

# TWILIGHT FACTOR OF VARIOUS BINOCULARS

| Size | TF | Size | TF | Size | TF |
|------|------|------|------|------|------|
| 5 x 20 | 10.0 | 6 x 30 | 13.4 | 10 x 50 | 22.4 |
| 6 x 20 | 11.0 | 7 x 30 | 14.5 | 12 x 50 | 24.5 |
| 7 x 20 | 11.8 | 8 x 30 | 15.5 | 16 x 50 | 28.3 |
| 8 x 20 | 12.6 | 9 x 30 | 16.4 | 20 x 50 | 31.6 |
| 9 x 20 | 13.4 | 10 x 30 | 17.3 | 21 x 50 | 32.4 |
| 10 x 20 | 14.1 | 6 x 32 | 13.9 | 24 x 50 | 34.6 |
| 6 x 21 | 11.2 | 8 x 32 | 16.0 | 35 x 50 | 41.8 |
| 7 x 21 | 12.1 | 10 x 32 | 17.9 | 8 x 56 | 21.2 |
| 8 x 21 | 13.0 | 6 x 35 | 14.5 | 10 x 56 | 23.7 |
| 10 x 21 | 14.5 | 7 x 35 | 15.7 | 10 x 60 | 24.5 |
| 8 x 22 | 13.3 | 9 x 35 | 17.7 | 15 x 60 | 30.0 |
| 6 x 23 | 11.7 | 10 x 35 | 18.7 | 20 x 60 | 34.6 |
| 8 x 23 | 13.6 | 12 x 35 | 20.5 | 30 x 60 | 42.4 |
| 10 x 23 | 15.2 | 15 x 35 | 22.9 | 9 x 63 | 23.8 |
| 12 x 23 | 16.6 | 17 x 35 | 24.4 | 12 x 63 | 27.5 |
| 6 x 24 | 12.0 | 6 x 36 | 14.7 | 10 x 70 | 26.5 |
| 7 x 24 | 13.0 | 8 x 36 | 17.0 | 14 x 70 | 31.3 |
| 8 x 24 | 13.9 | 12 x 36 | 20.8 | 20 x 70 | 37.4 |
| 9 x 24 | 14.7 | 7 x 40 | 16.7 | 30 x 70 | 45.8 |
| 10 x 24 | 15.5 | 8 x 40 | 17.9 | 11 x 80 | 29.7 |
| 12 x 24 | 17.0 | 9 x 40 | 19.0 | 15 x 80 | 34.6 |
| 15 x 24 | 19.0 | 10 x 40 | 20.0 | 16 x 80 | 35.8 |
| 6 x 25 | 12.2 | 14 x 40 | 23.7 | 20 x 80 | 40.0 |
| 7 x 25 | 13.2 | 16 x 40 | 25.3 | 30 x 80 | 49.0 |
| 8 x 25 | 14.1 | 21 x 40 | 29.0 | 14 x 100 | 37.4 |
| 9 x 25 | 15.0 | 7 x 42 | 17.1 | 20 x 100 | 44.7 |
| 10 x 25 | 15.8 | 7.5 x 42 | 17.7 | 25 x 100 | 50.0 |
| 12 x 25 | 17.3 | 8 x 42 | 18.3 | 20 x 120 | 49.0 |
| 15 x 25 | 19.4 | 10 x 42 | 20.5 | 25 x 150 | 61.2 |
| 7 x 26 | 13.5 | 7.5 x 44 | 18.2 | 35 x 150 | 72.5 |
| 10 x 26 | 16.1 | 7 x 50 | 18.7 | 40 x 150 | 77.5 |
| 10 x 28 | 16.7 | 8 x 50 | 20.0 | | |

The twilight factor is shown for the various sizes of binoculars including the low and high power of zoom binoculars

Resolution is a measurement of the binoculars' ability to distinguish fine detail (sharpness). Better resolution also provides more intense color.

Resolution varies directly with the size of the objective lens. All else being equal, a larger objective will always deliver more detail to the eye than a smaller objective lens regardless of the magnification of the binoculars.

It is measured in seconds of arc. The smaller the number of seconds of arc, the better the resolution. The human eye is believed to have a resolving power of 60 arc seconds.

One formula for determining the theoretical resolution of various size binoculars is to divide the objective lens diameter (in mm) into 116. For example, the resolution of 50mm diameter objectives is approximately 2.3 seconds of arc (116 divided by 50 equals 2.32). The accompanying table shows the resolution of various binoculars listed in this book.

$$\text{Resolution} = \frac{116}{\text{Objective Diameter (mm)}}$$

Keep in mind that these resolution numbers are theoretical. **Actual resolution is determined by the quality of the optical components, light transmission through the binoculars, atmospheric conditions, optical and mechanical alignment, collimation and the visual acuity of the individual.**

## (1) CONTRAST

Contrast is the degree to which both dim and bright objects in the image "stand out from" or "stand apart from" each other and from the general background. Higher contrast helps to see fainter objects or subtle detail which is important for deep space astronomy or for serious birders. Contrast is affected by the resolution such that the finer resolving power (in general), the better the contrast. The better the optical coatings, the better the contrast. Other factors that affect contrast are the quality of the optics (objectives, prisms and eyepieces), collimation and air turbulence.

# (2) OPTICAL PROBLEMS

There are several optical designs used for binoculars. Remember that a binocular is designed to gather more light than the unaided eye does and to enlarge the image of what you are looking at. In designing the optical system for a binocular, the optical engineer must make tradeoffs in controlling aberrations to achieve the desired result of the design.

Aberrations are any errors that result in the imperfection of an image. Such errors can result from design, fabrication or both. It is impossible to design an absolutely perfect binocular.

When you buy a binocular, your expectations are that it should be sharp at the center and also at the extreme edges. However, in reality this does not happen due to design compromises. Some aberrations of a minimal amount are present in all binoculars and generally are less apparent as the price of the binoculars increase. So, don't be too concerned unless the aberrations are severe or bothersome.

A brief description of aberrations in binoculars:

**Astigmatism** — This results from light rays focused as two lines perpendicular to each other rather than as a single point. It is detected by moving the binocular in and out of focus and watching stars at the edge of the field. If present, the focused lines go from vertical to horizontal and vice versa.

**Chromatic Aberration** (Color Aberration) — It is the failure to bring light of different wavelengths (colors) to a common focus. This results mainly in a faint colored halo (usually violet) around bright stars, the Moon and bright objects or dark objects against a bright sky. It also reduces contrast.

**Coma** — Generally caused by poor optical manufacturing. Coma affects the edges of the field of view and produces a V-shaped blurred image.

**Distortion** — It is generally caused when magnification varies from the center of the field to the edges. Looking at an object with straight lines (telephone wires, brick walls, etc.) that covers the entire field of view, you will notice a curving of the lines as you look near the edges of the field. If the lines are curved outward, this is called barrel distortion; and if the lines are curved inward, this is called pincushion distortion.

**Field Curvature** — This is caused by the light rays not all coming to a sharp focus in the same plane. The center of the field may be sharp and in focus but the edges are out of focus and vice versa.

**Spherical Aberration** — Here light rays at different distances from the optical center come to focus at different points on the axis. This causes a blurred image and you will not be able to focus sharply on any object.

For a more detailed discussion of optical aberrations for those of you who are interested, contact your local library for books on optical manufacturing.

**Again, don't be concerned with these potential problems. Usually, if any of them are severe, you will know it just by using them.**

| Objective Lens Diameter | Resolution In Arc Seconds | Objective Lens Diameter | Resolution In Arc Seconds |
|---|---|---|---|
| 20 | 5.8 | 40 | 2.9 |
| 21 | 5.5 | 42 | 2.8 |
| 22 | 5.3 | 44 | 2.6 |
| 23 | 5.0 | 50 | 2.3 |
| 24 | 4.8 | 56 | 2.1 |
| 25 | 4.6 | 60 | 1.9 |
| 26 | 4.5 | 63 | 1.8 |
| 28 | 4.1 | 70 | 1.7 |
| 30 | 3.9 | 80 | 1.5 |
| 32 | 3.6 | 100 | 1.2 |
| 35 | 3.3 | 120 | 1.0 |
| 36 | 3.2 | 150 | 0.8 |

The resolution is shown for the various size objective lenses (in millimeters) of binoculars rounded to the nearest tenth

**The biggest problem with binoculars that consumers have is poor collimation.** Collimation is the alignment of the optical elements of the binocular to the mechanical axis. Good collimation prevents eyestrain, headaches, inferior (or double) images and improves resolution.

Binoculars are normally tested for collimation by the manufacturer. Rough handling or shipping however can cause decollimation.

A high percentage of binoculars sold are out of collimation (mainly the inexpensive units). Most people don't realize this severe problem because their eyes try to compensate for the misalignment.

A couple of quick tests to determine if collimation is okay should be done: First focus on an object in the distance (like a street lamp or sign) about 100 yards or further away. Next have someone hold a book or other solid object over one objective lens. Now look through the binoculars (with both eyes open). Have the solid object removed from the one objective very quickly. If two images appear and then blend into one, the binoculars are out of collimation.

Another test is to look at the top of a brick or similar wall at a distance. Looking through the binoculars with both eyes, slowly pull them away from your eyes out to about 12 inches in front of you (keep looking through them). At first you see one view but as the binoculars are pulled away you will then see two views. These two views should be similar and in line. If one side or the other is higher or lower you have a collimation problem. If one side or the other is on top of the other, then you have a severe misalignment.

If you first look through a binocular and have double images, you need not do any other tests as obviously something is severely wrong.

**Center Focus** — most binoculars are of this type. Usually the right eyepiece has an individual eyesight adjustment collar (diopter). This permits you to compensate for any vision difference between your eyes.

Some manufacturers offer levers or other devices for obtaining focus faster but these usually require both hands to focus sharply and they may wear after time. So, I consider these devices as marketing hype and do not recommend them.

**Individual Focus** — allows extra-precise focusing adjustments for image sharpness and clarity. This type of focusing system is more reliable than center focusing. For subject matter 100 feet or more away with only one person using the binoculars, they are a good choice.

**Permanent Focus** — Jason Empire, in 1988, made a commercial success of binoculars that do not need focusing — they are, in essence, "permanently focused". Current units being sold are superior to the early models and now several manufacturers are offering this type of binocular. They are extremely easy to use. The negatives to this type of binocular (for most units) are that there is no way to adjust for vision differences in a user's eyes (eyeglass wearers must wear their glasses) and they do not have a close focusing range for birding and other applications. They are good for general viewing and sporting events.

Some companies with center and/or individual focusing systems say that once you focus at a certain distance (50 to 100 feet or so away), you now have permanently focused binoculars. They may be pretty close to being focused (as any binocular would be) but for critical use in observing objects at various distances you should refocus for the best sharpness.

## EYEPIECE FOCUSING METHODS

**External focus** — when turning the focusing wheel, the eyepieces move back and forth.

**Internal focus** — when turning the focusing wheel, you can't see anything move as the lenses inside the binoculars move.

Many people think internal focus is the best method since the binoculars are less exposed to contaminants. However, external focus is very good on most binoculars with the difference being insignificant.

**Body Covering** — this is a very subjective area. Many binoculars come with a leatherette-type covering. In recent years the rubberized armor covering has become more popular and these types withstand shock and rough handling a little better. More important than the covering is the appearance.

**Case** — most binoculars are supplied with a case. Generally, as the price of binoculars go up, the style of case and construction is better. When not using your binoculars, it is best to keep them in the case to keep them clean.

**Compass** — some binoculars are offered with compasses to find your location. They are useful for boaters, hunters, land viewing, etc.

**Comfort** — the ease of holding a binocular for long periods of time should be considered. Compare Porro prism versus roof prism designs yourself. Personally, I find Porro prisms more comfortable to hold and balance.

**Construction** — the security of the barrel alignment and the internal mounting of the optics is important to a long binocular life, free of user eye fatigue. It is difficult for you to determine how well a binocular is constructed but again, price is your best guideline; in general, the higher the price, the better the construction.

**Cosmetics** — binoculars come in all shapes, sizes and colors. You must decide on what is best for you.

**Filters** — usually they thread into the eyepiece (occasionally they thread into the objectives or fit over the objectives) for various special usages:

*LPR — city-bound astronomers will appreciate light pollution filters. They make the night sky appear darker while increasing the contrast of emission nebulae.*

*#12 Yellow — improves the visibility through haze and over water.*

*Solar — allows you to safely view the Sun for fantastic images of sunspots and granules.*

*Polarizing — cuts down on glare.*

**Numbered Reticles** — a ranging scale (reticle) is used to determine the distance you are from objects. These are good for boaters, hunters, military personnel, etc.

**Objective Lens and Eyepiece Caps** — again, most binoculars are supplied with these to prevent dust and other contaminants from accumulating on the optical elements when not in use.

**Straps** — most binoculars are supplied with a thin neck strap. In recent years many models are being supplied with a wider cloth strap which is much easier on your neck. If you have a very thin neck strap, consider buying an optional higher quality strap.

**Tripod Adaptability** — many binoculars have threads (1/4 x 20) in order to attach a binocular tripod adapter to allow you to mount the binoculars on a photographic/video tripod for more stable, vibration-free viewing. Be sure to use a stable tripod (not a $50 or so unit) or you will not achieve the purpose of using the tripod — stability!

But don't worry if your binoculars don't have the threads built in as some firms such as Innovative Energies Inc. in Newton, Massachusetts now offer special products to allow your binoculars to be adapted to a tripod – the only drawback being that it will cost you more money than the conventional tripod adapter.

**Warranty** — most binoculars have a long warranty with many for life. Stick with a reputable manufacturer as a lifetime warranty from a "fly-by-night" firm will do you no good if you can't get them serviced or the company is no longer in business.

**Weight** — this is an important aspect to consider when comparing binoculars. Most binoculars offered today are considerably lighter in weight than those offered a decade ago.

Photo showing a high quality neck strap

# CHAPTER 8

## A. INTERPUPILLARY (INTEROCULAR) DISTANCE

Since the distance between the eyes (specifically the distance between the centers of the pupils) varies among individuals, the two eyepieces must be correctly aligned. This is called adjusting the interpupillary distance.

To adjust this distance, begin by setting the diopter eyepiece to 0. Lift the binoculars up and look through them. Move the two halves of the binocular about the hinge until you see one clear circle of image. Now you are ready to observe.

**If your binoculars have an interpupillary scale on the hinge, note the setting and it will be faster to set up when using the binoculars again.**

## B. ADJUSTING FOCUS

Since most people have a variance in vision from their left eye to their right, you must adjust the focusing system.

*Center Focus Binoculars* - use the following steps to achieve focus: (1) close your right eye and look through the left side of the binoculars with your left eye. Rotate the center focusing wheel until the image appears in sharp focus; (2) next, close your left eye and look through the right eyepiece. Rotate the diopter eyepiece until the image appears in sharp focus; (3) now look through both eyepieces with both eyes open. Since you've already adjusted the right eyepiece, use only the center focusing wheel to refocus on a new object at a different distance.

Some experts advise covering the objective lens of each side with your hand while performing the above focusing adjustment. This is okay but I feel the closing of the eye method is best as the binoculars are more stable when doing the adjustment because both hands are holding the binocular steady.

*Individual Focus Binoculars* - you must close one eye at a time (it doesn't make any difference which one is first) and rotate the eyepiece until the image is in sharp focus. When changing distances of various objects observed, you must refocus each eyepiece.

*Permanent Focus* - there is no focus adjustment.

Photo showing a typical diopter adjustment scale on a binocular

Photo showing interpupillary scale

# C. INSPECTION, CLEANING AND REPAIR

When you first buy a pair of binoculars, check the overall appearance. Make sure the outside of the unit is clean and free of scratches or dents. Check the optical appearance (objective lenses, eyepieces, prisms and coatings) for any scratches, a lot of dirt or other obvious defects. Inspect the focusing mechanism for smoothness. Then look through the binoculars at an object to check optical quality. You should be able to focus sharply and feel eye comfort. Do not look through windows and expect the same sharpness as you would outdoors. Most windows are of poor optical glass which is especially noticeable when you are looking out at an angle.

To verify optical quality, there are a few things you can look at to help you and determine if you do have any severe aberrations (as discussed earlier in this book).

Look at a brick wall or other object that has straight lines and fills the entire field of view – check the straightness of the lines from the center to the edge of the field. Then check for sharpness at the center and the edge.

Now look at a sign, chart or decorative building outside in daylight that is brightly lit – check for sharpness, brightness and natural colors.

You can look at a chart or other object with fine detail in the darkened corner of a store or look at an object outside at twilight to ascertain the brightness and detail available.

The most rigid of tests to look for aberrations is on a star.

Binoculars do not need routine maintenance other than making sure the objectives and eyepieces are clean. If repairs become necessary, they should be serviced by the manufacturer or a qualified binocular repair company.

As noted earlier in this book, collimation is the biggest concern with binoculars. If your binoculars are roughly handled or dropped, there is a good chance that the collimation will be out and they should be serviced.

Dirty objectives lenses and/or eyepieces mean less light transmission and loss of brightness as well as unsharp images. Keep your optics clean! When not using your binoculars, keep the lens caps on and store them in a case. Avoid touching the glass surfaces but if fingerprints (which can contain mild acid) get on them, they should be cleaned as soon as possible to avoid damaging the coatings. To clean the optical surfaces, I recommend a lens/optics cleaning kit available at most optical suppliers. If you have multi-coatings there are special cleaning kits especially made for these. If you have a lot of dust or dirt accumulated, brush it off gently with a camel's hair brush and/or utilize a can of pressurized air. Then use the cleaning kit.

## D. ADAPTING BINOCULARS TO A TRIPOD

To mount this adapter, remove the hinge cap and thread the adapter screw into the hinge socket and tighten. The other end of the adapter mounts to the pan head of the tripod.

Photo showing a binocular tripod adapter and where it is inserted on most binoculars

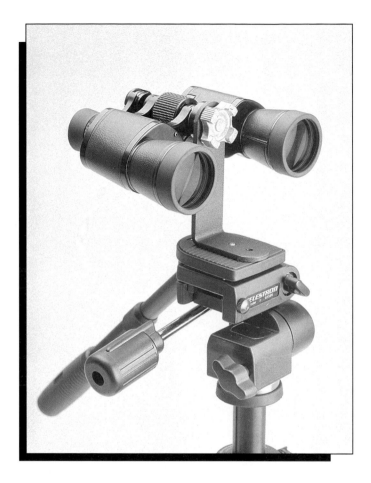

Photo showing binoculars adapted to a photo
tripod by usage of a binocular tripod adapter

# CHAPTER 9

## WATERPROOF BINOCULARS

True waterproof binoculars are nitrogen purged (filled) internally and O-ring sealed. To qualify as waterproof, most Japanese manufacturers have established that units should withstand submersion in approximately 16 feet of water for up to five minutes without damage.

These units are impervious to saltwater and dust. The nitrogen prevents oxidation and internal fogging.

They are great for boating or other wet weather activities.

I laugh at all the various terms manufacturers use for binoculars that are water-resistant — *they are not waterproof.* Here's a sampling; moisture resistant, splashproof, sprayproof, fogproof, rainproof, weatherproof and finally, showerproof.

Celestron 8 - 24 x 50 Pro zoom binocular
Photo courtesy of Celestron International

# CHAPTER 10

Knowledgeable optical experts usually reject the idea of using zoom binoculars. However, many consumers demand zoom binoculars because it gives them a convenient range of magnifications in one package.

In general, zoom binocular optical performance is not as good as that of similarly priced standard (one fixed power) binoculars. They often have to be refocused after zooming. Also, high zoom magnifications become increasingly more difficult to hold steady and are bigger, heavier and more costly than conventional binoculars. Zoom binocular eyepiece designs are usually compromised and have narrower fields than fixed powers.

On the positive side, they are convenient to use and when more power is needed for particular subject matter, you just turn the lever. Or on some models an electric zooming mechanism is offered.

The zooming mechanism should be predictable, smooth and easy to operate. Two basic mechanisms are used: (1) the belt system; and, (2) the gear system. The gear system is more reliable and performs better over a long period of time.

**A tip on focusing zoom binoculars is to always focus at the higher power first.**

If you decide to purchase a zoom binocular, try to obtain one with a large objective lens (40 to 50mm) as the image will be brighter at the high power range as compared to smaller objective lenses.

# CHAPTER 11

An exception to using prisms are the Galilean binoculars which are called "sport glasses" or "opera glasses". These are simple, compact devices utilizing only lenses with no need for erecting prisms.

This type of binocular has a magnification range of 2.5 to 5x with objective lens sizes from 12 to 30mm, which is good for concerts, the theater or indoor sporting events, etc., where most conventional binoculars are too powerful. Higher magnifications and a wide field of view cannot be obtained with this design. Good features are that they are rugged, lightweight and compact. Also, they are good for children as they are more durable since they lack a prism assembly that could come out of alignment.

Very few manufacturers offer opera glasses which is a shame since they are very useful. A combined listing of opera glasses can be found in the tables at the rear of this book.

# CHAPTER 12

A monocular is one-half of a binocular. It is much lighter and more compact than a binocular. Monoculars are very useful for the same applications as binoculars and should be considered for their versatility especially for hiking, backpacking and other activities where light weight is very important.

I consider a monocular to be in the range of 20 to 30mm for the objective lenses. Smaller than 20mm is just too small to be very useful and over 30mm should be considered a spotting scope.

Not too many manufacturers offer monoculars. So, a combined listing of monoculars can be found in the tables at the rear of this book.

There are no waterproof monoculars offered to my knowledge. All monoculars have a near focus in the range of ten to twenty feet.

There are even specialized monoculars offered to aid golfers in their game. These monoculars have a built-in distance scale to determine the distance to the pin.

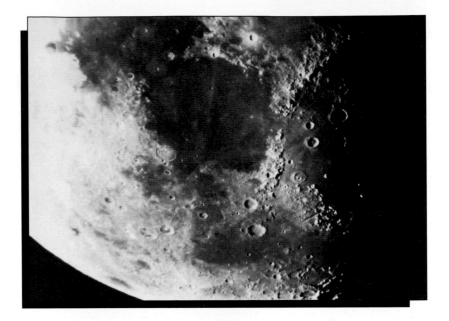

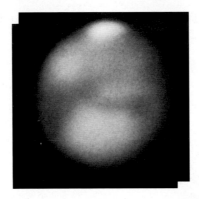

# CHAPTER 13

You may purchase your binoculars initially for a particular usage. But once you have the basic tool, you may want to expand your horizons. The brief list of applications at the beginning of the book can give you ideas.

Our world is a fantastic place to continually enjoy. With increased concern for our environment, take a closer look with binoculars at what you have been missing.

If you are not a birder, consider starting out casually. Casual birding can be fun! Most of us have had an interest in birds since we were children. We all look at birds but so much more can be revealed when looking at them with binoculars. All you need are your binoculars and a field guide of birds. A wealth of information is available on birds in book stores and specialty nationwide outlets such as The Nature Company and others.

Another interesting hobby is astronomy. I believe that all humans have a latent interest in astronomy. No, you don't necessarily need a telescope to enjoy the heavens. With binoculars you can enjoy looking at the planets (you can see Jupiter and some of its moons and the rings of Saturn), study the Moon, observe star clusters, comets, nebulae and galaxies. Simple star charts will show you where to look and you'll find it's easy and fun. In contrast to the numerous books on birding, there are few books on using binoculars for astronomy. The better books include:

"Astronomy With Binoculars" by James Muirden

"Exploring The Night Sky With Binoculars" by Patrick Moore

"Star Gazing Through Binoculars" by Stephen Mensing

"Touring The Universe Through Binoculars" by Philip Harrington

# CHAPTER 14

Binoculars can be purchased from numerous sources. Outlets include:

- photo/camera stores
- sporting good stores
- electronic stores
- department stores
- outdoor and nature stores
- catalog showrooms
- museums
- astronomy stores
- mail-order catalogs

# CHAPTER 15

## WHAT NEW TECHNOLOGY OR IMPROVEMENTS CAN BE COMING?

Binoculars have been in existence for a long period of time. Product changes have come about slowly evolving toward better quality offerings.

In late-1990, Minolta was the first company to offer true auto focusing binoculars with the same technology used in their sophisticated cameras. I'm sure others will follow shortly.

Look for better and improved optical coatings as techniques and methods are improved.

More and more binoculars will be offered with longer eye relief to help eyeglass wearers.

There is always ongoing research to find better materials for the glass used in objective lenses, eyepieces and prisms.

To meet consumer desires, there will always be ongoing changes in cosmetic appearance, not only in style and colors, but in weight and "feel" of the binoculars.

Possibly some totally new technology will make its appearance that we don't even conceive of today.

# CHAPTER 16

Aus Jena c/o Europtik, Ltd.
P.O. Box 319
Dunmore, PA 18512
(717) 347-6049

Bausch & Lomb
c/o Bushnell

Brunton U.S.A.
620 E. Monroe Ave.
Riverton, WY 82501
(307) 856-6559

Bushnell - Div of Bausch & Lomb
300 N. Lone Hill Ave.
San Dimas, CA 91773
(714) 592-8072

Celestron International
2835 Columbia Street
Torrance, CA 90503
(213) 328-9560
(800) 421-1526

Fujinon Inc.
10 High Point Dr.
Wayne, NJ 07470
(201) 633-5600

Jason Empire
920 Cody
Overland Park, KS 66214
(913) 888-0220

Kowa Optimed, Inc.
20001 S. Vermont Ave.
Torrance, CA 90502
(213) 327-1913

Leica Camera Inc.
156 Ludlow Avenue
Northvale, NJ 07647
(201) 767-7500

Leupold & Stevens, Inc.
P.O. Box 688
Beaverton, OR 97075
(503) 646-9171

Minolta Corporation
101 William Drive
Ramsey, NJ 07446
(201) 825-4000

Mirador Optical Corporation
P.O. Box 11614
Marina del Rey, CA 90295
(213) 821-5587

Nikon Inc.
1300 Walt Whitman Road
Melville, NY 11747
(516) 547-4200

Optolyth U.S.A.
18805 N.E. Melvista Lane
Hillsboro, OR 97123
(503) 628-0246

Orion Telescope Center
2450 17th Avenue
Santa Cruz, CA 95061
(800) 447-1001
(800) 443-1001 CA

Parks Optical
270 Easy Street
Simi Valley, CA 93065
(805) 522-6722

Pentax
35 Iverness Dr., E.
Englewood, CO 80112
(303) 799-8000

Redfield Inc.
5800 E. Jewell Avenue
Denver, CO 80224
(303) 757-6411

Selsi Company, Inc.
40 Veterans Blvd.
Carlstadt, NJ 07072
(201) 935-0388

Simmons Outdoor Corporation
14530 S.W. 119th Avenue
Miami, FL 33186
(305) 252-0477

Steiner c/o Pioneer Marketing
& Research, Inc.
216 Haddon Avenue, Suite 500
Westmont, NJ 08108
(609) 854-2424

Swarovski Optik
2 Slater Road
Cranston, RI 02920
(401) 463-3000
(800) 556-6478

Swift Instruments Inc.
952 Dorchester Avenue
Boston, MA 02125
(617) 436-2960
(800) 446-1116

Tasco
7600 N.W. 26th Street
Miami, FL 33122
(305) 591-3670

Vixen Optical Industries Ltd.
c/o Celestron International

Carl Zeiss Optical, Inc.
1015 Commerce Street
Petersburg, VA 23803
(804) 862-3734
(800) 338-2984

# CHAPTER 17

- Aus Jena — Limited Lifetime
- Brunton — 5 Years
- Bushnell — Limited Lifetime
  (Bausch & Lomb)
- Celestron — Limited Lifetime
- Fujinon — Limited Lifetime
- Jason — ?
- Kowa — Limited Lifetime
- Leica — 3 Years Unconditional &
  then Limited Lifetime
- Leupold — Limited Lifetime
- Minolta — 25 Years
  (Except Auto Focus - 5 Years)
- Mirador — Limited Lifetime
- Nikon — 25 Years
- Optolyth — Limited Lifetime
- Orion — 2 Years
- Parks — Limited Lifetime
- Pentax — Limited Lifetime
- Redfield — Limited Lifetime
- Selsi — ?
- Simmons — Limited Lifetime
  (Except Red Line - 1 Year)
- Steiner — 5 Years
- Swarovski — Limited Lifetime
- Swift — Limited Lifetime
- Tasco — Limited Lifetime
- Vixen — 5 Years
- Zeiss — ?

# CHAPTER 18

## Using The Tables

**Size** — Power and Objective Lens Diameter

**Mfg** — Manufacturer

**Series** — Manufacturer's Designation

**Model** — Manufacturer's Model Callout

**FOV** — Angular Field of View (in Degrees), For Zoom Binoculars the FOV at the Lowest Power

**RA** — Rubber Armor - Y (Yes) or N (No)

**ER** — Eye Relief in Millimeters

**PR** — Prisms Used - PP-4 (for Porro Prisms BAK-4) - PP-7 (for Porro Prisms BK-7) - RP (for Roof Prism)

**CO** — Type of Optical Coatings - C (Coated), FC (Fully Coated), MC (Multi-Coated), FMC (Fully Multi-Coated)

**NF** — Near Focus in Feet

**TOF** — Type of Focus - CF (Center Focus), IF (Individual Focus), FF (Fixed Focus), AF(Auto Focus)

**WP** — Waterproof - Y (Yes) or N (No)

**CA** — Case Supplied - Y (Yes) or N (No)

**S** — Straps for the Binoculars Supplied - Y (Yes) or N (No)

**CPS** — Caps or Covers Supplied for *Both* Eyepieces and Objectives - Y (Yes) or N (No)

**TA** — Tripod Adaptable - Y (Yes) or N (No)

**WT** — Weight in Ounces

**SL** — Suggested List Price (latest known prices as of April 1991)

**?** — Question mark means that the manufacturer does not or would not supply this information

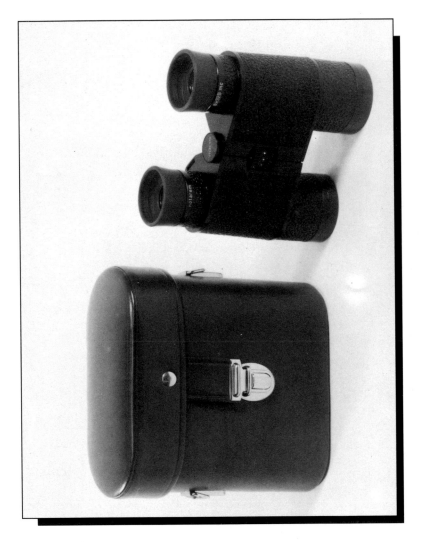

Aus Jena 8x32 Notarem
Photo courtesy of Europtik, Ltd.

# AUS JENA

| SIZE | SERIES | MODEL | FOV | RA | ER | PR | CO | NF | TOF | WP | CA | S | CPS | TA | WT | SL |
|---|---|---|---|---|---|---|---|---|---|---|---|---|---|---|---|---|
| 8 x 30 | WA | TRADITIONAL | 8.5 | N | ? | PP-4 | MC | 7 | CF | N | Y | Y | N | Y | 18 | $510 |
| 7 X 50 | ? | TRADITIONAL | 7.3 | N | ? | PP-4 | MC | 14 | CF | N | Y | Y | N | Y | 36 | 590 |
| 10 X 50 | WA | TRADITIONAL | 7.3 | N | ? | PP-4 | MC | 14 | CF | N | Y | Y | N | Y | 36 | 610 |
| 10 X 40 | B | NOTAREM | 6 | N | ? | RP | MC | 17 | CF | N | Y | Y | Y | N | 22 | 695 |
| 10 X 40 | BGA | NOTAREM | 6 | Y | ? | RP | MC | 17 | CF | N | N | Y | Y | N | 22 | 720 |
| 8 X 32 | B | NOTAREM | 7.4 | N | ? | RP | MC | 12 | CF | N | Y | Y | Y | N | 18 | 665 |
| 8 X 32 | BGA | NOTAREM | 7.4 | Y | ? | RP | MC | 12 | CF | N | N | Y | N | N | 18 | 680 |
| 8 X 50 | B | OCTAREM | 7.6 | N | ? | PP-4 | MC | 12 | CF | N | Y | Y | Y | Y | 38 | 755 |
| 8 X 50 | BGA | OCTAREM | 7.6 | Y | ? | PP-4 | MC | 12 | CF | N | N | Y | N | Y | 45 | 770 |
| 12 X 50 | B | DODECAREM | 5.6 | N | ? | PP-4 | MC | 15 | CF | N | Y | Y | Y | Y | 39 | 780 |
| 12 X 50 | BGA | DODECAREM | 5.6 | Y | ? | PP-4 | MC | 15 | CF | N | N | Y | N | Y | 47 | 795 |
| 7 X 40 | B | SAFARI | 7.5 | Y | 20 | RP | MC | 14 | IF | N | N | Y | Y | N | 35 | 725 |
| 7 X 50 | B | NOBILEM | 8.5 | N | ? | PP-4 | MC | ? | CF | N | Y | Y | Y | Y | 39 | 740 |
| 7 X 50 | BGA | NOBILEM | 8.5 | Y | ? | PP-4 | MC | ? | CF | N | N | Y | N | Y | ? | 755 |
| 8 X 56 | B | NOBILEM | 7.6 | N | ? | PP-4 | MC | ? | CF | N | Y | Y | Y | Y | 44 | 810 |
| 10 X 50 | B | NOBILEM | 8 | N | ? | PP-4 | MC | ? | CF | N | Y | Y | Y | Y | 40 | 765 |
| 10 X 50 | BGA | NOBILEM | 8 | Y | ? | PP-4 | MC | ? | CF | N | N | Y | N | Y | ? | 780 |
| 15 X 60 | B | NOBILEM | 4.9 | N | ? | PP-4 | MC | ? | CF | N | Y | Y | Y | Y | 52 | 995 |

Bausch & Lomb Discoverer 7x35 (9x35 similar appearance)
Photo courtesy of Bushnell

# BAUSCH & LOMB

| SIZE | SERIES | MODEL | FOV | RA | ER | PR | CO | NF | TOF | WP | CA | S | CPS | TA | WT | SL |
|---|---|---|---|---|---|---|---|---|---|---|---|---|---|---|---|---|
| 8 X 42 | ELITE | 61-2843 | 7 | Y | 20 | RP | FMC | 12 | CF | N | Y | Y | N | N | 28 | $1810 |
| 10 X 42 | ELITE | 61-1042 | 5.6 | Y | 20 | RP | FMC | 12 | CF | N | Y | Y | N | N | 28 | 1863 |
| 7 X 35 | DISCOVERER | 61-2010 | 7.3 | N | 9-12 | PP-4 | FMC | 15 | CF | N | Y | Y | N | N | 22 | 634 |
| 9 X 35 | DISCOVERER | 61-2030 | 7.3 | N | 9-12 | PP-4 | FMC | 15 | CF | N | Y | Y | N | N | 23 | 674 |
| 8 X 36 | CUSTOM | 61-3360 | 6.5 | N | 19 | PP-4 | FMC | 13 | CF | N | Y | Y | N | N | 23 | 465 |
| 10 X 40 | CUSTOM | 61-1046 | 5.2 | Y | 19 | PP-4 | FMC | 7-10 | CF | N | Y | Y | N | N | 29 | 508 |
| 7 X 26 | CUSTOM | 61-7261 | 7.4 | N | 16 | PP-4 | FMC | 10 | CF | N | Y | Y | N | N | 11 | 439 |
| 7 X 35 | LEGACY | 12-7355 | 11 | Y | 9-12 | PP-4 | MC | 15 | CF | N | Y | Y | N | N | 28 | 166 |
| 10 X 50 | LEGACY | 12-1055 | 7.9 | Y | 9-12 | PP-4 | MC | 15 | CF | N | Y | Y | N | Y | 35 | 212 |
| 7-15 X 35 | LEGACY | 12-7135 | 5.6@7x | Y | 9-12 | PP-4 | MC | 15-20 | CF | N | Y | Y | N | Y | 26 | 284 |
| 8-24 X 50 | LEGACY | 12-3246 | 4.7@8x | Y | 9-12 | PP-4 | MC | 15-20 | CF | N | Y | Y | N | Y | 30 | 317 |
| 9 X 25 | LEGACY | 12-0925 | 7.3 | Y | 9-12 | PP-4 | MC | 15 | CF | N | Y | Y | N | N | 10 | 160 |
| 7-15 X 25 | LEGACY | 12-7125 | 4.9@7x | Y | 9-12 | PP-4 | MC | 15-20 | CF | N | Y | Y | N | Y | 13 | 252 |
| 8 X 20 | LEGACY | 12-8205 | 7 | Y | 9-12 | RP | MC | 15 | CF | N | Y | Y | N | N | 7 | 200 |
| 8 X 24 | LEGACY | 12-8240 | 6.6 | N | 9-12 | PP-4 | MC | ? | CF | N | Y | Y | N | N | 9 | 159 |
| 11 X 80 | DISCOVERER | 11-1180 | 4.2 | N | 9-12 | PP-4 | MC | 30 | CF | N | Y | Y | N | Y | 80 | 727 |
| 20 X 80 | DISCOVERER | 11-2080 | 2.7 | N | 9-12 | PP-4 | MC | 40 | CF | N | Y | Y | N | Y | 80 | 758 |

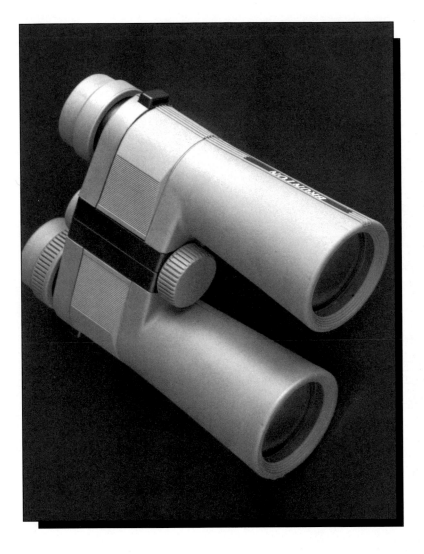

Brunton 7x42 (10x42 similar) Tracker II
Photo courtesy of Brunton U.S.A.

96

# BRUNTON

| SIZE | SERIES | MODEL | FOV | RA | ER | PR | CO | NF | TOF | WP | CA | S | CPS | TA | WT | SL |
|------|--------|-------|-----|----|----|----|----|----|-----|----|----|---|-----|----|----|----|
| 8 X 21 | SUPER BANTAM | 4030 | 6 | Y | ? | RP | FC | 20 | CF | N | Y | Y | N | N | 8 | $179 |
| 10 X 24 | GRAY HAWK | 4050 | 5 | Y | ? | RP | FC | 23 | CF | N | Y | Y | N | N | 12 | 188 |
| 7 X 42 | TRACKER II | 4055 | 7.1 | Y | ? | RP | FC | 10 | CF | N | Y | Y | Y | N | 24 | 374 |
| 10 X 42 | TRACKER III | 4056 | 5 | Y | ? | RP | FC | 10 | CF | N | Y | Y | Y | N | 24 | 399 |
| 8 X 25 | LITE TECH | 4057 | 8.2 | Y | ? | RP | FC | 10 | CF | N | Y | Y | Y | N | 9 | 119 |
| 10 X 25 | LITE TECH | 4058 | 5.5 | Y | ? | RP | FC | 20 | CF | N | Y | Y | N | N | 10 | 121 |
| 8 X 25 | PASSPORT I | 4054 | 6 | Y | ? | RP | FC | 23 | CF | Y | Y | Y | N | N | 13 | 200 |
| 10 X 25 | PASSPORT II | 4059 | 6.5 | Y | ? | RP | FC | 36 | CF | Y | Y | Y | N | N | 13 | 210 |

Bushnell Spectator 7-21x40 Zoom
Photo courtesy of Bushnell

# BUSHNELL

| SIZE | SERIES | MODEL | FOV | RA | ER | PR | CO | NF | TOF | WP | CA | S | CPS | TA | WT | SL |
|---|---|---|---|---|---|---|---|---|---|---|---|---|---|---|---|---|
| 7 X 35 | SPECTATOR | 13-7735 | 9.5 | Y | 9-12 | PP-7 | FC | 15 | CF | N | Y | Y | N | N | 17 | $125 |
| 7 X 50 | SPECTATOR | 13-7750 | 7.1 | Y | 9-12 | PP-7 | FC | 15 | CF | N | Y | Y | N | N | 26 | 158 |
| 8 X 40 | SPECTATOR | 13-7784 | 8.2 | Y | 9-12 | PP-7 | FC | 15 | CF | N | Y | Y | N | N | 23 | 150 |
| 10 X 50 | SPECTATOR | 13-7710 | 7 | Y | 9-12 | PP-7 | FC | 15 | CF | N | Y | Y | N | N | 26 | 161 |
| 8 X 42 | SPECTATOR | 13-8042 | 6.8 | Y | 9-12 | RP | FC | 15 | CF | N | Y | Y | N | N | 22 | 378 |
| 7-15 X 35 | SPECTATOR | 13-7136 | 5.7@7X | Y | 9-12 | PP-7 | FC | ? | CF | N | Y | Y | N | N | 22 | 204 |
| 7-21X 40 | SPECTATOR | 13-7215 | 5.7@7X | Y | 9-12 | PP-7 | FC | ? | CF | N | Y | Y | N | N | 24 | 243 |
| 8 X 23 | SPECTATOR | 13-3230 | 7 | Y | 9-12 | PP-7 | FC | 15 | CF | N | Y | Y | N | N | 9 | 113 |
| 8 X 21 | SPECTATOR | 13-3212 | 7 | Y | 9-12 | RP | FC | 15 | CF | N | Y | Y | N | N | 8 | 161 |
| 10 X 25 | SPECTATOR | 13-2512 | 6.5 | Y | 9-12 | RP | FC | 15 | CF | N | Y | Y | N | N | 9 | 169 |
| 8 X 25 | SPECTATOR | 13-3253 | 8.2 | Y | 9-12 | RP | FC | 15 | CF | Y | Y | Y | N | N | 12 | 319 |
| 7 X 50 | SPECTATOR | 13-7552 | 7.5 | Y | 9-12 | PP-7 | FC | 15 | CF | Y | Y | Y | N | N | 39 | 361 |
| 7 X 35 | SPECTATOR | 13-3350 | 7 | Y | 9-12 | PP-7 | FC | 15 | FF | N | Y | Y | N | N | 19 | 113 |
| 10 X 50 | SPECTATOR | 13-3360 | 5 | Y | 9-12 | PP-7 | FC | 20 | FF | N | Y | Y | N | N | 29 | 146 |
| 7 X 50 | ENSIGN | 13-7551 | 7.1 | Y | 9-12 | PP-7 | FC | ? | CF | N | Y | Y | N | N | 29 | 123 |
| 10 X 25 | ENSIGN | 13-2511 | 6.5 | N | 9-12 | RP | FC | ? | CF | N | Y | Y | N | N | 9 | 133 |
| 7-15 X 35 | ENSIGN | 13-7015 | 5.7@7X | N | 9-12 | PP-7 | FC | ? | CF | N | Y | Y | N | N | 22 | 134 |
| 7-21 X 40 | ENSIGN | 13-7214 | 5.5@7X | N | 9-12 | PP-7 | FC | ? | CF | N | Y | Y | N | N | 24 | 161 |
| 7 X 50 | MARINE | 12-8755 | 7.5 | Y | 9-12 | PP-4 | FC | ? | IF | Y | Y | Y | N | N | 37 | 421 |
| 7 X 35 | BIRDER | 11-7305 | 6.8 | N | ? | PP-? | ? | ? | CF | N | Y | Y | N | N | 21 | 84 |

100

Celestron 11x80 Giant
Photo courtesy of Celestron International

# CELESTRON

| SIZE | SERIES | MODEL | FOV | RA | ER | PR | CO | NF | TOF | WP | CA | S | CPS | TA | WT | SL |
|---|---|---|---|---|---|---|---|---|---|---|---|---|---|---|---|---|
| 8 X 22 | MINI | 71124 | 8.2 | Y | 14 | RP | MC | 10 | CF | N | Y | Y | Y | N | 9 | $160 |
| 8 X 25 | MINI | 71503 | 8.7 | Y | 5.3 | RP | MC | 15 | CF | N | Y | Y | Y | N | 9 | 110 |
| 10 X 25 | MINI | 71125 | 6.5 | Y | 11 | RP | MC | 12 | CF | N | Y | Y | Y | N | 9 | 170 |
| 8 X 40 | FIXED FOCUS | 71128 | 8.5 | Y | 12 | PP-7 | FC | 53 | FF | N | Y | Y | Y | Y | 22 | 120 |
| 7 X 50 | FIXED FOCUS | 71129 | 7 | Y | 21 | PP-7 | FC | 40 | FF | N | Y | Y | Y | Y | 27 | 120 |
| 10 X 50 | FIXED FOCUS | 71130 | 7 | Y | 13 | PP-7 | FC | 82 | FF | N | Y | Y | Y | Y | 27 | 120 |
| 8 X 30 | PRO | 71121 | 8.5 | Y | 10 | PP-4 | FMC | 15 | CF | N | Y | Y | Y | Y | 22 | 200 |
| 7 X 50 | PRO | 71122 | 7.1 | Y | 19 | PP-4 | FMC | 25 | CF | N | Y | Y | Y | Y | 28 | 220 |
| 10 X 50 | PRO | 71123 | 6.5 | Y | 13 | PP-4 | FMC | 35 | CF | N | Y | Y | Y | Y | 29 | 230 |
| 12 X 50 | PRO | 71120 | 5.5 | Y | 10 | PP-4 | FMC | 23 | CF | N | Y | Y | Y | Y | 29 | 250 |
| 8-24 X 50 | PRO | 71119 | 5@8X | Y | 5 | PP-4 | MC | 28 | CF | N | Y | Y | Y | Y | 31 | 350 |
| 8 X 32 | ULTIMA | 71110 | 8.3 | N | 15 | PP-4 | FMC | 11 | CF | N | Y | Y | Y | Y | 18 | 330 |
| 7 X 42 | ULTIMA | 71109 | 7 | N | 23 | PP-4 | FMC | 21 | CF | N | Y | Y | Y | Y | 22 | 340 |
| 10 X 42 | ULTIMA | 71108 | 6.6 | N | 11 | PP-4 | FMC | 19 | CF | N | Y | Y | Y | Y | 20 | 350 |
| 10 X 50 | ULTIMA | 71127 | 5 | N | 19 | PP-4 | FMC | 36 | CF | N | Y | Y | Y | Y | 26 | 370 |
| 8 X 56 | ULTIMA | 71126 | 6.1 | N | 17 | PP-4 | FMC | 39 | CF | N | Y | Y | Y | Y | 30 | 400 |
| 8 X 40 | SPORT | 71140 | 8.5 | Y | 12 | PP-7 | FC | 20 | CF | N | Y | Y | Y | Y | 24 | 120 |
| 7 X 50 | SPORT | 71141 | 7 | Y | 21 | PP-7 | FC | 21 | CF | N | Y | Y | Y | Y | 29 | 130 |
| 10 X 50 | SPORT | 71142 | 7 | Y | 13 | PP-7 | FC | 30 | CF | N | Y | Y | Y | Y | 29 | 130 |
| 8 X 21 | WATERPROOF | 71507 | 6 | Y | 15 | RP | FMC | 18 | CF | Y | Y | Y | Y | N | 14 | 340 |
| 8 X 30 | WATERPROOF | 71511 | 8.3 | Y | 10 | RP | MC | 20 | IF | Y | Y | Y | Y | N | 23 | 590 |
| 10 X 40 | WATERPROOF | 71510 | 6.5 | Y | 11 | RP | MC | 25 | IF | Y | Y | Y | Y | Y | 27 | 620 |
| 7 X 50 | WATERPROOF | 71508 | 7.5 | Y | 16.5 | PP-4 | FMC | 15 | CF | Y | Y | Y | Y | Y | 39 | 360 |
| 7 X 50 | WATERPROOF-M | 71512 | 7.3 | Y | 12 | PP-4 | MC | 36 | IF | Y | Y | Y | Y | Y | 51 | 700 |
| 10 X 50 | WATERPROOF | 71509 | 6.5 | Y | 9 | PP-4 | FMC | 25 | CF | Y | Y | Y | Y | Y | 39 | 370 |
| 11 X 80 | GIANT | 71105 | 4.5 | N | 16 | PP-4 | FMC | 41 | CF | N | Y | Y | Y | Y | 80 | 590 |
| 15 X 80 | GIANT | 71106 | 3.5 | N | 17 | PP-4 | FMC | 66 | CF | N | Y | Y | Y | Y | 80 | 600 |
| 20 X 80 | GIANT | 71107 | 3.5 | N | 14 | PP-4 | FMC | 53 | CF | N | Y | Y | Y | Y | 80 | 620 |

Fujinon 6x30 Nautilus
Photo courtesy of Fujinon Inc.

# FUJINON INC.

| SIZE | SERIES | MODEL | FOV | RA | ER | PR | CO | NF | TOF | WP | CA | S | CPS | TA | WT | SL |
|---|---|---|---|---|---|---|---|---|---|---|---|---|---|---|---|---|
| 7 X 35 | 2000 | 7207350 | 11 | Y | 14 | PP-? | C | ? | CF | N | Y | Y | Y | Y | 26 | $143 |
| 8 X 40 | 2000 | 7208400 | 9 | Y | 14 | PP-? | C | ? | CF | N | Y | Y | Y | Y | 29 | 165 |
| 10 X 50 | 2000 | 7210500 | 7.5 | Y | 14 | PP-? | C | ? | CF | N | Y | Y | Y | Y | 34 | 176 |
| 7-21 X 40 | 2000 | 7272140 | 5.3@7X | Y | 15 | PP-? | C | ? | CF | N | Y | Y | Y | Y | 31 | 220 |
| 7 X 42 | 4000 | 7407420 | 6.5 | N | 12 | RP | MC | ? | CF | N | Y | Y | Y | Y | 19 | 230 |
| 8 X 32 | 4000 | 7408320 | 7.5 | N | 14 | RP | MC | ? | CF | N | Y | Y | Y | Y | 17 | 230 |
| 10 X 42 | 4000 | 7410420 | 5.5 | N | 12 | RP | MC | ? | CF | N | Y | Y | Y | Y | 19 | 240 |
| 7 X 50 | NAUTILUS A | 7107509 | 7.5 | Y | 16 | PP-? | FC | ? | IF | Y | N | Y | Y | Y | 46 | 370 |
| 7 X 50 | POSEIDON SX | 7107501 | 7.5 | N | 12 | PP-? | FMC | ? | IF | Y | Y | Y | Y | Y | 45 | 550 |
| 7 X 50 | POSEIDON SX | 7107506 | 7.5 | Y | 12 | PP-? | FMC | ? | IF | Y | N | Y | Y | Y | 48 | 600 |
| 10 X 70 | POSEIDON SX | 7110701 | 5.3 | N | 12 | PP-? | FMC | ? | IF | Y | Y | Y | Y | Y | 74 | 720 |
| 14 X 70 | POSEIDON SX | 7114701 | 4 | N | 12 | PP-? | FMC | ? | IF | Y | Y | Y | Y | Y | 72 | 750 |
| 7 X 50 | POLARIS F-SX | 7107502 | 7.5 | N | 23 | PP-? | FMC | ? | IF | Y | Y | Y | Y | Y | 50 | 680 |
| 7 X 50 | POLARIS F-SX | 7107507 | 7.5 | Y | 23 | PP-? | FMC | ? | IF | Y | N | Y | Y | Y | 53 | 730 |
| 10 X 70 | POLARIS F-SX | 7107702 | 5.3 | N | 23 | PP-? | FMC | ? | IF | Y | Y | Y | Y | Y | 76 | 850 |
| 8 X 30 | POSEIDON MT | 7108300 | 7.5 | N | 14 | RP | FC | ? | IF | Y | Y | Y | Y | Y | 24 | 410 |
| 8 X 30 | POSEIDON MT | 7108305 | 7.5 | Y | 14 | RP | FC | ? | IF | Y | Y | Y | Y | Y | 25 | 460 |
| 10 X 32 | ROOF PRISM | 7010320 | 5.5 | N | 14 | RP | FMC | ? | CF | N | Y | N | Y | Y | 15 | 310 |
| 12 X 36 | ROOF PRISM | 7012360 | 4.6 | N | 14 | RP | FMC | ? | CF | N | Y | N | Y | Y | 18 | 340 |
| 15 X 80 | BINOCULAR TELESCOPE | 7115800 | 4 | N | 15 | PP-? | FMC | ? | IF | Y | Y | N | Y | Y | 247 | 2650 |
| 25 X 150 | BINOCULAR TELESCOPE | 7125150 | 2.7 | N | 18 | PP-7 | FMC | ? | IF | Y | Y | N | Y | Y | 1005 | 7200 |
| 6 X 30 | NAUTILUS | 7106300 | 8 | Y | 16 | PP-4 | FC | ? | IF | Y | Y | Y | Y | Y | 26 | 280 |
| 6 X 30 | POLARIS F-SX | 7107908 | 8.3 | Y | 16 | PP-? | FMC | ? | IF | Y | Y | Y | Y | Y | 25 | 600 |

# JASON

| SIZE | SERIES | MODEL | FOV | RA | ER | PR | CO | NF | TOF | WP | CA | S | CPS | TA | WT | SL |
|---|---|---|---|---|---|---|---|---|---|---|---|---|---|---|---|---|
| 7 X 40 | EAGLE | 173 | 9.3 | ? | ? | PP-4 | ? | ? | CF | N | ? | ? | ? | ? | 23 | $218 |
| 8 X 42 | EAGLE | 174 | 8.2 | ? | ? | PP-4 | ? | ? | CF | N | ? | ? | ? | ? | 23 | 235 |
| 10 X 56 | EAGLE | 175 | 6.5 | ? | ? | PP-4 | ? | ? | CF | N | ? | ? | ? | ? | 23 | 327 |
| 7-21 X 40 | STATESMAN | 177 | 5.5@7X | N | ? | PP-4 | FC | ? | CF | N | Y | ? | ? | ? | 25 | 218 |
| 8-24 X 50 | STATESMAN | 178 | 4.6@8X | N | ? | PP-4 | FC | ? | CF | N | Y | ? | ? | ? | 29 | 228 |
| 10 X 50 | MERCURY | 1142 | 7 | N | ? | PP-? | ? | ? | CF | N | Y | ? | ? | ? | 32 | 123 |
| 7 X 50 | ALL TERRAIN | 114 | 7.1 | Y | ? | PP-? | FC | ? | CF | N | Y | ? | ? | ? | 31 | 143 |
| 10 X 50 | ALL TERRAIN | 115 | 7 | Y | ? | PP-? | FC | ? | CF | N | Y | ? | ? | ? | 32 | 147 |
| 8 X 21 | ALL TERRAIN | 117 | 7 | Y | ? | RP | FC | ? | CF | N | Y | ? | ? | ? | 8 | 143 |
| 10 X 25 | ALL TERRAIN | 120 | 5.5 | Y | ? | RP | FC | ? | CF | N | Y | ? | ? | ? | 10 | 153 |
| 7 X 35 | PERMA FOCUS 2000 | 194 | 6.5 | Y | ? | PP-? | FC | 40 | FF | N | Y | Y | ? | ? | 18 | 122 |
| 7 X 35 | PERMA FOCUS 2000 | 195 | 11 | Y | ? | PP-? | FC | 40 | FF | N | Y | Y | ? | ? | 23 | 147 |
| 10 X 50 | PERMA FOCUS 2000 | 196 | 7.5 | Y | ? | PP-? | FC | 90 | FF | N | Y | Y | ? | ? | 30 | 160 |
| 8 X 40 | PERMA FOCUS 2000 | 201 | 6 | Y | ? | PP-? | FC | 40 | FF | N | Y | Y | ? | ? | 20 | 141 |
| 7 X 42 | PERMA FOCUS 2000 | 209 | 6.5 | Y | ? | RP | FC | 40 | FF | N | Y | Y | ? | ? | 17 | 296 |
| 7 X 50 | PERMA FOCUS 2000 | 200 | 7.1 | Y | ? | PP-? | FC | 40 | FF | N | Y | Y | ? | ? | 30 | 155 |
| 6-12 X 35 | PERMA FOCUS 2000 | 188 | 5.8@6X | Y | ? | PP-? | FC | 40 | FF | N | Y | Y | ? | ? | 22 | 167 |
| 6 X 21 | PERMA FOCUS 2000 | 184 | 7.7 | N | ? | PP-? | FC | 40 | FF | N | Y | Y | ? | ? | 5 | 109 |
| 7 X 21 | PERMA FOCUS 2000 | 190 | 7.5 | Y | ? | RP | FC | 40 | FF | N | Y | Y | ? | ? | 7 | 142 |
| 7 X 25 | PERMA FOCUS 2000 | 193 | 9.3 | Y | ? | PP-? | FC | 40 | FF | N | Y | Y | ? | ? | 16 | 132 |
| 7 X 50 | PERMA FOCUS 2000 | 212 | 7.1 | Y | ? | RP | FC | 40 | FF | Y | Y | Y | ? | ? | 29 | 354 |
| 8 X 25 | PERMA FOCUS 2000 | 215 | 8.2 | Y | ? | PP-? | FC | 40 | FF | Y | Y | Y | ? | ? | 11 | 226 |
| 7 X 42 | PERMA FOCUS 2000 | 98-PF | 6.5 | N | ? | PP-? | FC | 40 | FF | Y | Y | Y | ? | ? | 22 | 343 |

# KOWA

| SIZE | SERIES | MODEL | FOV | RA | ER | PR | CO | NF | TOF | WP | CA | S | CPS | TA | WT | SL |
|---|---|---|---|---|---|---|---|---|---|---|---|---|---|---|---|---|
| 10 X 42 | LUPINUS | ? | 5 | Y | 17 | RP | MC | ? | CF | N | Y | ? | ? | ? | 33 | $428 |
| 8 X 42 | LUPINUS | ? | 6.2 | Y | 17 | RP | MC | ? | CF | N | Y | ? | ? | ? | 33 | 358 |
| 15 X 80 | OBSERVATION | BL-8E | 4 | N | ? | PP-? | ? | ? | IF | Y | ? | ? | Y | ? | ? | ? |
| 20 X 80 | OBSERVATION | BL-8G | 3 | N | ? | PP-? | ? | ? | IF | Y | ? | ? | Y | ? | ? | ? |
| 20 X 120 | OBSERVATION | BL-12C | 3 | N | ? | PP-? | ? | ? | IF | N | ? | ? | ? | ? | ? | ? |
| 35 X 150 | OBSERVATION | BL-15C | 1.7 | N | ? | PP-? | ? | ? | IF | N | ? | ? | ? | ? | ? | ? |
| 40 X 150 | OBSERVATION | BL-15D | 1.5 | N | ? | PP-? | ? | ? | IF | N | ? | ? | ? | ? | ? | ? |

Leica Ultra-Line Binoculars
Photo courtesy of Leica Camera Inc.

# LEICA

| SIZE | SERIES | MODEL | FOV | RA | ER | PR | CO | NF | TOF | WP | CA | S | CPS | TA | WT | SL |
|---|---|---|---|---|---|---|---|---|---|---|---|---|---|---|---|---|
| 8 X 42 | BA ULTRA | 040-012 | 7.4 | Y | 15.9 | RP | FMC | 17 | CF | Y | Y | Y | N | N | 31 | $1365 |
| 7 X 42 | BA ULTRA | 040-013 | 8 | Y | 17 | RP | FMC | 21 | CF | Y | Y | Y | N | N | 31 | 1365 |
| 10 X 42 | BA ULTRA | 040-014 | 6.3 | Y | 13.9 | RP | FMC | 15 | CF | Y | Y | Y | N | N | 31 | 1380 |
| 8 X 20 | BCA TRINOVID | 040-309 | 6.6 | Y | 13.9 | RP | FC | 7 | CF | N | Y | Y | N | N | 8 | 615 |
| 10 X 25 | BCA TRINOVID | 040-310 | 5.4 | Y | 13.7 | RP | FC | 10 | CF | N | Y | Y | N | N | 9 | 645 |
| 10 X 42 | B ULTRA | 040-229 | 6.3 | N | 13.9 | RP | FMC | 15 | CF | Y | Y | Y | N | N | 31 | 1380 |

# LEUPOLD

| SIZE | SERIES | MODEL | FOV | RA | ER | PR | CO | NF | TOF | WP | CA | S | CPS | TA | WT | SL |
|------|--------|-------|-----|----|----|----|----|----|-----|----|----|----|-----|----|----|----|
| 7 X 20 | POCKET | 41200 | 6.8 | N | 13.8 | RP | FC | 15 | IF | N | Y | Y | N | N | 10 | $391 |
| 9 X 25 | POCKET | 41202 | 5.3 | N | 13.8 | RP | FC | 15 | IF | N | Y | Y | N | N | 11 | 427 |
| 9 X 35 | M-2 | 43318 | 7.3 | N | 11.6 | PP-4 | FC | 15 | IF | N | Y | Y | N | Y | 21 | 355 |
| 10 X 40 | M-2 | 43319 | 6.6 | N | 11.4 | PP-4 | FC | 15 | IF | N | Y | Y | N | Y | 23 | 373 |
| 7 X 20 | POCKET | 41201 | 6.8 | Y | 13.8 | RP | FC | 15 | IF | N | Y | Y | N | N | 10 | 420 |
| 9 X 25 | POCKET | 41203 | 5.3 | Y | 13.8 | RP | FC | 15 | IF | N | Y | Y | N | N | 11 | 463 |

Minolta Pocket Line of Binoculars
Photo courtesy of Minolta Corporation

# MINOLTA

| SIZE | SERIES | MODEL | FOV | RA | ER | PR | CO | NF | TOF | WP | CA | S | CPS | TA | WT | SL |
|------|--------|-------|-----|----|----|----|----|----|-----|----|----|----|-----|----|----|----|
| 7 X 42 | WEATHERMATIC | 8556-107 | 9.3 | Y | 13 | RF | MC | 20 | CF | Y | Y | Y | Y | Y | 30 | $387 |
| 7 X 50 | WEATHERMATIC | 8557-107 | 7.1 | Y | 15.1 | PP-4 | MC | 33 | CF | Y | Y | Y | Y | Y | 33 | 364 |
| 10 X 42 | WEATHERMATIC | 8556-207 | 6.5 | Y | 8.8 | RP | MC | 33 | CF | Y | Y | Y | Y | Y | 30 | 455 |
| 7 X 35 | STANDARD | 8554-117 | 11 | Y | 12 | PP-4 | MC | 20 | CF | N | Y | Y | Y | Y | 28 | 134 |
| 7 X 50 | STANDARD | 8554-607 | 7.8 | Y | 18 | PP-4 | MC | 30 | CF | N | Y | Y | Y | Y | 33 | 182 |
| 8 X 40 | STANDARD | 8554-217 | 9.5 | Y | 11.1 | PP-4 | MC | 26 | CF | N | Y | Y | Y | Y | 35 | 148 |
| 10 X 50 | STANDARD | 8554-317 | 7.8 | Y | 11.1 | PP-4 | MC | 33 | CF | N | Y | Y | Y | Y | 35 | 194 |
| 7-15 X 35 | STANDARD | 8554-417 | 5.5@7X | Y | 12 | PP-4 | MC | 30 | CF | N | Y | Y | Y | Y | 29 | 235 |
| 7-21 X 50 | STANDARD | 8554-517 | 5@7X | Y | 7 | PP-4 | MC | 36 | CF | N | Y | Y | Y | Y | 36 | 296 |
| 7 X 21 | POCKET | 8555-107 | 7.3 | Y | 10.6 | RP | MC | 6 | CF | N | Y | Y | Y | N | 11 | 130 |
| 8 X 22 | POCKET | 8555-207 | 8.2 | Y | 8 | RP | MC | 11 | CF | N | Y | Y | Y | N | 11 | 153 |
| 9 X 24 | POCKET | 8555-307 | 5.7 | Y | 10.9 | RP | MC | 16 | CF | N | Y | Y | Y | N | 11 | 166 |
| 10 X 25 | POCKET | 8555-407 | 6.5 | Y | 8 | RP | MC | 17 | CF | N | Y | Y | Y | N | 11 | 189 |
| 8 X 23 | COMPACT | 8559-117 | 7 | N | 10 | PP-4 | MC | 13 | CF | N | Y | Y | Y | N | 8 | 126 |
| 10 X 23 | COMPACT | 8559-217 | 5 | N | 12.2 | PP-4 | MC | 20 | CF | N | Y | Y | Y | N | 8 | 159 |
| 8 X 24 | COMPACT | 8553-217 | 7 | N | 7.2 | PP-4 | MC | 13 | CF | N | Y | Y | Y | N | 11 | 159 |
| 10 X 25 | COMPACT | 8553-317 | 6.3 | N | 7 | PP-4 | MC | 17 | CF | N | Y | Y | Y | N | 11 | 182 |
| 8 X 22 | AUTO FOCUS | 8563-113 | 6.5 | N | 7.5 | RP | MC | 7 | AF | N | Y | Y | N | N | 19 | 376 |
| 10 X 25 | AUTO FOCUS | 8563-213 | 5.2 | N | 7.5 | RP | MC | 9 | AF | N | Y | Y | N | N | 19 | 421 |

# MIRADOR

| SIZE | SERIES | MODEL | FOV | RA | ER | PR | CO | NF | TOF | WP | CA | S | CPS | TA | WT | SL |
|---|---|---|---|---|---|---|---|---|---|---|---|---|---|---|---|---|
| 7 X 35 | EXL | 7350 | 7.3 | N | 11.8 | PP-4 | MC | ? | CF | N | Y | Y | Y | Y | 22 | $459 |
| 9 X 35 | EXL | 9350 | 7.3 | N | 10 | PP-4 | MC | ? | CF | N | Y | Y | Y | Y | 22 | 504 |
| 7 X 50 | EXL | 7500 | 7.3 | N | 13 | PP-4 | MC | ? | CF | N | Y | Y | Y | Y | 46 | 549 |
| 10 X 50 | EXL | 1050 | 7 | N | 13.3 | PP-4 | MC | ? | CF | N | Y | Y | Y | Y | 46 | 585 |
| 6 X 32 | YCC | 6320 | 8.2 | N | 19.3 | PP-4 | MC | 12 | CF | N | Y | Y | Y | Y | 17 | 297 |
| 8 X 32 | YCC | 8320 | 8.2 | N | 11.8 | PP-4 | MC | 11 | CF | N | Y | Y | Y | Y | 18 | 306 |
| 7 X 42 | YCC | 7420 | 6.6 | N | 22 | PP-4 | MC | 21 | CF | N | Y | Y | Y | Y | 22 | 315 |
| 8 X 42 | YCC | 8420 | 6.3 | N | 18 | PP-4 | MC | 18 | CF | N | Y | Y | Y | Y | 22 | 324 |
| 10 X 42 | YCC | 1042 | 6.6 | N | 11.4 | PP-4 | MC | 16 | CF | N | Y | Y | Y | Y | 22 | 333 |
| 10 X 50 | YCC | 1050 | 5 | N | 18.3 | PP-4 | MC | 29 | CF | N | Y | Y | Y | Y | 26 | 369 |
| 6 X 30 | ACC | 6300 | 8.2 | N | 19.3 | PP-4 | MC | 11 | CF | N | Y | Y | Y | Y | 20 | 360 |
| 7 X 35 | ACC | 7350 | 7.3 | N | 19 | PP-4 | MC | 13 | CF | N | Y | Y | Y | Y | 21 | 369 |
| 7X 40 | ACC | 7400 | 6.6 | N | 22 | PP-4 | MC | 21 | CF | N | Y | Y | Y | Y | 24 | 378 |
| 8 X 40 | ACC | 8400 | 6.3 | N | 18.7 | PP-4 | MC | 18 | CF | N | Y | Y | Y | Y | 24 | 387 |
| 9 X 35 | ACC | 9350 | 7.3 | N | 11.6 | PP-4 | MC | 13 | CF | N | Y | Y | Y | Y | 21 | 387 |
| 10 X 40 | ACC | 1040 | 6.6 | N | 11.4 | PP-4 | MC | 16 | CF | N | Y | Y | Y | Y | 24 | 396 |
| 8 X 30 | DIF-WP | 8308 | 8.5 | Y | ? | RP | MC | ? | IF | Y | N | Y | Y | N | 24 | 495 |
| 7 X 35 | BCF | 7350 | 9.5 | N | ? | PP-4 | FC | ? | CF | N | Y | Y | Y | Y | 19 | 144 |
| 7 X 50 | BCF | 7500 | 7.2 | N | ? | PP-4 | FC | ? | CF | N | Y | Y | Y | Y | 25 | 171 |
| 10 X 50 | BCF | 1050 | 7 | N | ? | PP-4 | FC | ? | CF | N | Y | Y | Y | Y | 25 | 171 |
| 11 X 80 | GIANT | 1´80 | 4.5 | N | ? | PP-4 | MC | ? | CF | N | Y | Y | Y | Y | 61 | 612 |
| 20 X 80 | GIANT | 2080 | 3.5 | N | ? | PP-4 | MC | ? | CF | N | Y | Y | Y | Y | 61 | 639 |
| 8 X 24 | KSL | 8240 | 5.5 | N | ? | PP-4 | FC | ? | CF | N | Y | Y | N | N | 13 | 144 |
| 8 X 21 | DCF | 8210 | 7 | N | ? | RP | FC | ? | CF | N | Y | Y | N | N | 7 | 144 |
| 8 X 21 | DCF | 8214 | 7 | Y | ? | RP | FC | ? | CF | N | Y | Y | N | N | 10 | 153 |

Nikon 6-12x24 Le Tour Zoom
Photo courtesy of Nikon Inc.

# NIKON

| SIZE | SERIES | MODEL | FOV | RA | ER | PR | CO | NF | TOF | WP | CA | S | CPS | TA | WT | SL |
|---|---|---|---|---|---|---|---|---|---|---|---|---|---|---|---|---|
| 7 X 20 | TRAVELITE III | ? | 7.1 | N | 10.2 | PP-4 | FMC | ? | CF | N | Y | Y | N | N | 7 | $118 |
| 9 X 25 | TRAVELITE III | ? | 5.6 | N | 10.2 | PP-4 | FMC | ? | CF | N | Y | Y | N | N | 9 | 140 |
| 8 X 23 | VENTURER II | 7842 | 6.3 | N | 10.2 | PP-4 | FMC | ? | CF | N | Y | Y | N | N | 9 | 132 |
| 10 X 25 | VENTURER II | 7843 | 5 | N | 9.2 | PP-4 | FMC | ? | CF | N | Y | Y | N | N | 10 | 160 |
| 6-12 X 24 | LE TOUR | 7861 | 6.9@6X | Y | 17.6 | RP | FC | ? | CF | N | Y | Y | Y | N | 21 | 598 |
| 8 X 20 | LE TOUR | 7827 | 6.5 | Y | 9.7 | RP | FC | ? | CF | N | Y | Y | Y | N | 7 | 303 |
| 10 X 25 | LE TOUR | 7828 | 5.2 | Y | 9.7 | RP | FC | ? | CF | N | Y | Y | Y | N | 9 | 333 |
| 7 X 35 | STAY FOCUS PLUS | 7873 | 8.6 | N | 9 | PP-4 | FMC | 40 | FF | N | Y | Y | Y | N | 21 | 150 |
| 7 X 50 | STAY FOCUS PLUS | 7874 | 6.2 | N | 16.4 | PP-4 | FMC | 40 | FF | N | Y | Y | Y | N | 30 | 166 |
| 10 X 50 | STAY FOCUS PLUS | 7875 | 6 | N | 9 | PP-4 | FMC | 80 | FF | N | Y | Y | Y | N | 30 | 180 |
| 8-16 X 40 | ZOOM XL | 7860 | 5.2@8X | N | 12 | PP-4 | FC | ? | CF | N | Y | Y | Y | N | 31 | 920 |
| 9 X 30 | EXECULITE | 711 | 6.7 | N | 11.1 | RP | FC | ? | CF | N | Y | Y | Y | N | 16 | 540 |
| 12 X 36 | EXECULITE | 714 | 5 | N | 12.5 | RP | FC | ? | CF | N | Y | Y | Y | N | 20 | 583 |
| 10 X 70 | ASTROLUXE | 7788 | 5.1 | N | 15 | PP-4 | FMC | ? | IF | Y | Y | Y | Y | Y | 88 | 1015 |
| 7 X 50 | PROSTAR | 7789 | 7.3 | N | 15 | PP-4 | FMC | ? | IF | Y | Y | Y | Y | Y | 52 | 1058 |
| 8 X 40 | CLASSIC EAGLE | 7826 | 7 | N | 16.4 | RP | FMC | ? | CF | Y | Y | Y | Y | N | 30 | 1107 |
| 7 X 35 | E | 780 | 7.3 | N | 16 | PP-4 | FMC | ? | CF | N | Y | Y | Y | N | 22 | 420 |
| 8 X 30 | E | 782 | 8.3 | N | 13 | PP-4 | FMC | ? | CF | N | Y | Y | Y | N | 20 | 460 |
| 10 X 35 | E | 784 | 6.6 | N | 12.4 | PP-4 | FMC | ? | CF | N | Y | Y | Y | N | 22 | 500 |
| 10 X 25 | MOUNTAINEER | 7845 | 5 | Y | 10.4 | PP-4 | FMC | ? | CF | Y | Y | Y | N | N | 17 | 350 |
| 8 X 23 | MOUNTAINEER | 7846 | 6.3 | Y | 10.6 | PP-4 | FMC | ? | CF | Y | Y | Y | N | N | 15 | 324 |
| 7 X 50 | SEAFARER II | ? | 7.3 | Y | 15 | PP-4 | FC | ? | IF | Y | N | Y | Y | N | 61 | 625 |
| 8 X 30 | TRAILBLAZER | 7782 | 7.5 | Y | 13.2 | RP | FC | ? | IF | Y | Y | Y | Y | N | 24 | 519 |
| 7 X 50 | SEASIDER | 7866 | 6.2 | Y | 16.4 | PP-4 | FMC | ? | IF | Y | Y | Y | Y | N | 38 | 320 |
| 7 X 50 | WINDJAMMER | 7865 | 6.2 | Y | 16.4 | PP-4 | FMC | ? | IF | Y | Y | Y | Y | N | 38 | 320 |
| 7-15 X 35 | SCOUTMASTER | 7863 | 5.8@7X | N | 12 | PP-4 | FMC | ? | CF | N | Y | Y | N | N | 25 | 199 |
| 7 X 35 | NATURALIST II | 7870 | 8.6 | N | 9 | PP-4 | FMC | ? | CF | N | Y | Y | Y | N | 20 | 128 |
| 10 X 50 | LOOKOUT II | 7872 | 6 | N | 9 | PP-4 | FMC | ? | CF | N | Y | Y | Y | N | 29 | 165 |

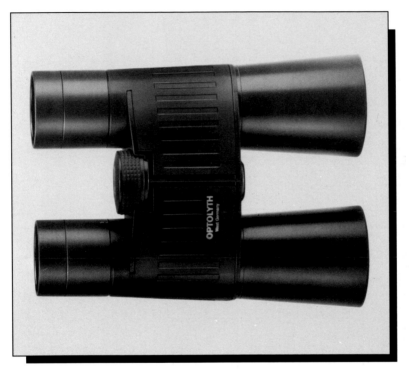

Optolyth 10x40 Touring
Photo courtesy of Optolyth U.S.A.

**116**

# OPTOLYTH

| SIZE | SERIES | MODEL | FOV | RA | ER | PR | CO | NF | TOF | WP | CA | S | CPS | TA | WT | SL |
|---|---|---|---|---|---|---|---|---|---|---|---|---|---|---|---|---|
| 8 X 22 | SPORTING | 15010 | 6.7 | N | 14 | RP | FMC | 10 | CF | N | Y | ? | Y | N | 6 | $445 |
| 10 X 25 | SPORTING | 15020 | 5.8 | N | 14 | RP | FMC | 13 | CF | N | Y | ? | Y | N | 7 | 475 |
| 8 X 32 | TOURING | 13010 | 7.7 | Y | 10 | RP | FMC | 12 | CF | N | Y | Y | Y | N | 22 | 850 |
| 7 X 42 | TOURING | 13020 | 8 | Y | 10 | RP | FMC | 22 | CF | N | Y | Y | Y | N | 30 | 950 |
| 10 X 40 | TOURING | 13030 | 6.3 | Y | 11 | RP | FMC | 20 | CF | N | Y | Y | Y | N | 25 | 918 |
| 8 X 30 | ALPIN | 12160 | 8 | Y | 10 | PP-4 | FMC | 12 | CF | N | Y | Y | Y | N | 15 | 445 |
| 8 X 40 | ALPIN | 12170 | 6.6 | Y | 16 | PP-4 | FMC | 16 | CF | N | Y | Y | Y | N | 17 | 500 |
| 10 X 40 | ALPIN | 12180 | 6.6 | Y | 10 | PP-4 | FMC | 13 | CF | N | Y | Y | Y | N | 17 | 550 |
| 7 X 42 | ALPIN | 12190 | 6.6 | Y | 16 | PP-4 | FMC | 16 | CF | N | Y | Y | Y | N | 17 | 550 |
| 7 X 50 | ALPIN | 12200 | 5.6 | Y | 16 | PP-4 | FMC | 17 | CF | N | Y | Y | Y | N | 23 | 600 |
| 10 X 50 | ALPIN | 12210 | 5.1 | Y | 16 | PP-4 | FMC | 19 | CF | N | Y | Y | Y | N | 23 | 610 |
| 12 X 50 | ALPIN | 12220 | 5 | Y | 10 | PP-4 | FMC | 19 | CF | N | Y | Y | Y | N | 23 | 620 |
| 7 X 50 | ROYAL | 14151 | 7.7 | Y | ? | RP | FMC | ? | CF | N | Y | Y | Y | N | 36 | 1300 |
| 8 X 56 | ROYAL | 14021 | 6.7 | Y | 10 | RP | FMC | 12 | CF | N | Y | Y | Y | N | 39 | 1230 |
| 9 X 63 | ROYAL | 14071 | 6.1 | Y | ? | RP | FMC | ? | CF | N | Y | Y | Y | N | 42 | 1435 |
| 12 X 63 | ROYAL | 14161 | 5 | Y | 5 | RP | FMC | 15 | CF | N | Y | Y | Y | Y | 42 | 1500 |

Orion 9x63 Little Giant
Photo courtesy of Orion Telescope Center

# ORION

| SIZE | SERIES | MODEL | FOV | RA | ER | PR | CO | NF | TOF | WP | CA | S | CPS | TA | WT | SL |
|------|--------|-------|-----|----|----|----|----|----|-----|----|----|----|-----|----|----|-----|
| 11 X 80 | GIANT | 9502 | 4.5 | N | 16 | PP-4 | FMC | 66 | CF | N | Y | Y | Y | Y | 81 | $349 |
| 16 X 80 | GIANT | 9496 | 3.5 | N | 15.5 | PP-4 | FMC | 82 | CF | N | Y | Y | Y | Y | 77 | 379 |
| 20 X 80 | GIANT | 9498 | 3.5 | N | 15 | PP-4 | FMC | 66 | CF | N | Y | Y | Y | Y | 77 | 389 |
| 14 X 100 | SUPER GIAN" | 9495 | 3.3 | N | 24 | PP-4 | MC | 82 | IF | N | Y | N | Y | Y | 112 | 799 |
| 8 X 56 | SKY EXPLORER | 9504 | 6 | N | 18 | PP-4 | MC | 36 | CF | N | Y | Y | Y | Y | 33 | 139 |
| 9 X 63 | LITTLE GIANT | 9500 | 5.5 | Y | 13 | RP | FMC | 30 | CF | N | Y | Y | Y | Y | 42 | 199 |
| 10 X 70 | GIANT | 9501 | 5 | N | 15 | PP-4 | FMC | 82 | CF | N | Y | Y | Y | Y | 56 | 279 |
| 7 X 50 | OBSERVER | 9620 | 7 | Y | 17 | PP-7 | FC | 26 | CF | N | Y | Y | Y | Y | 28 | 85 |
| 8 X 40 | OBSERVER | 9621 | 8.5 | Y | 12 | PP-7 | FC | 26 | CF | N | Y | Y | Y | Y | 23 | 80 |
| 8 X 30 | OBSERVER | 9622 | 8.5 | Y | 10 | PP-7 | FC | 26 | CF | N | Y | Y | Y | Y | 21 | 75 |
| 8 X 21 | SUPER COMPACT | 9499 | 7 | Y | 10 | RP | FC | 13 | CF | N | Y | Y | N | N | 6 | 80 |

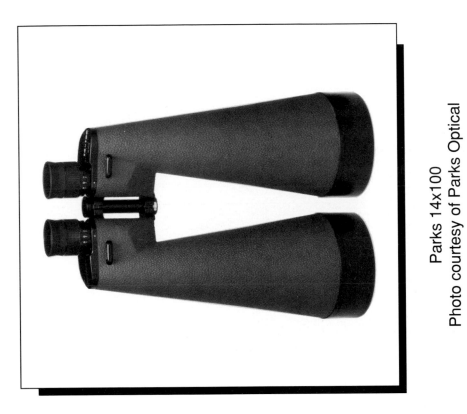

Parks 14x100
Photo courtesy of Parks Optical

# PARKS

| SIZE | SERIES | MODEL | FOV | RA | ER | PR | CO | NF | TOF | WP | CA | S | CPS | TA | WT | SL |
|---|---|---|---|---|---|---|---|---|---|---|---|---|---|---|---|---|
| 11 X 80 | | 401-30000 | 4.5 | Y | ? | PP-2 | MC | 30 | CF | N | Y | Y | Y | Y | 80 | $700 |
| 15 X 80 | | 401-30010 | 3.5 | Y | ? | PP-2 | MC | 45 | CF | N | Y | Y | Y | Y | 80 | 700 |
| 20 X 80 | | 401-30020 | 3.5 | Y | ? | PP-2 | MC | 50 | CF | N | Y | Y | Y | Y | 80 | 700 |
| 14 X 100 | | 401-30100 | 3.3 | N | ? | PP-4 | MC | ? | IF | N | Y | Y | Y | Y | ? | 1850 |
| 25 X 100 | | 401-30110 | 2.6 | N | ? | PP-4 | MC | ? | CF | N | Y | Y | Y | Y | ? | 1850 |
| 7 X 35 | | 401-20000 | 7 | N | ? | PP-4 | MC | ? | CF | N | Y | Y | Y | N | 22 | 150 |
| 8 X 42 | | 401-20030 | 6.5 | Y | ? | RP | MC | ? | CF | N | Y | Y | Y | N | 22 | 400 |
| 9-17 X 35 | | 401-20040 | ? | Y | ? | RP | MC | ? | CF | N | Y | Y | Y | N | 20 | 400 |
| 6 X 36 | | 401-20050 | ? | Y | ? | ? | MC | ? | FF | N | Y | Y | Y | ? | 18 | 300 |
| 8 X 25 | | 401-10010 | 8.7 | Y | ? | RP | MC | ? | CF | N | Y | Y | N | N | ? | 175 |
| 12 X 63 | | 401-20060 | 5 | Y | ? | ? | MC | ? | CF | N | Y | Y | Y | Y | ? | 400 |

# PENTAX

| SIZE | SERIES | MODEL | FOV | RA | ER | PR | CO | NF | TOF | WP | CA | S | CPS | TA | WT | SL |
|---|---|---|---|---|---|---|---|---|---|---|---|---|---|---|---|---|
| 7 X 20 | DCF | 62320 | 7.5 | Y | 12 | RP | MC | 8 | CF | N | Y | Y | N | N | 7 | $200 |
| 7 X 21 | DCF | 61310 | 7.5 | Y | 15 | RP | MC | 13 | CF | N | Y | Y | N | N | 13 | 320 |
| 8 X 42 | DCF | 62340 | 6.2 | Y | 17 | RP | MC | 20 | CF | N | Y | Y | Y | N | 31 | 375 |
| 8 X 56 | DCF | 62360 | 6.5 | N | 21.5 | RP | MC | 26 | CF | N | Y | Y | Y | Y | 41 | 710 |
| 9 X 20 | DCF | 62330 | 6.2 | N | 12 | RP | MC | 8 | CF | N | Y | Y | N | N | 7 | 220 |
| 9 X 63 | DCF | 62370 | 5.5 | N | 21.5 | RP | MC | 33 | CF | N | Y | Y | Y | Y | 44 | 780 |
| 10 X 42 | DCF | 62350 | 5 | N | 17 | RP | MC | 20 | CF | N | Y | Y | Y | N | 31 | 420 |
| 7 X 35 | PCF | 65600 | 9.3 | Y | 14 | PP-4 | MC | 20 | CF | N | Y | Y | Y | Y | 27 | 200 |
| 7 X 50 | PCF | 65610 | 7.1 | Y | 20 | PP-4 | MC | 20 | CF | N | Y | Y | Y | Y | 32 | 220 |
| 8 X 40 | PCF | 65620 | 8.2 | Y | 14 | PP-4 | MC | 20 | CF | N | Y | Y | Y | Y | 27 | 215 |
| 10 X 50 | PCF | 65650 | 6.5 | Y | 14 | PP-4 | MC | 20 | CF | N | Y | Y | Y | Y | 32 | 225 |
| 12 X 50 | PCF | 65660 | 5.5 | Y | 14 | PP-4 | MC | 20 | CF | N | Y | Y | Y | Y | 32 | 235 |
| 16 X 50 | PCF | 65670 | 3 | Y | 14 | PP-4 | MC | 30 | CF | N | Y | Y | Y | Y | 32 | 255 |
| 8 X 24 | UCF | 62130 | 7.5 | Y | 12 | PP-4 | MC | 13 | CF | N | Y | Y | N | N | 9 | 160 |
| 10 X 24 | UCF | 62131 | 6 | Y | 10 | PP-4 | MC | 13 | CF | N | Y | Y | N | N | 9 | 170 |
| 12 X 24 | UCF | 62132 | 5 | Y | 8 | PP-4 | MC | 13 | CF | N | Y | Y | N | N | 9 | 180 |

123

Redfield 7x35 and 10x50 Waterproof Models
Photo courtesy of Redfield Inc.

124

# REDFIELD

| SIZE | SERIES | MODEL | FOV | RA | ER | PR | CO | NF | TOF | WP | CA | S | CPS | TA | WT | SL |
|------|--------|-------|-----|-----|-----|-----|-----|-----|------|-----|-----|-----|-----|-----|-----|-----|
| 7 X 35 | WP-ADJUST-FREE | 235000 | 7.3 | Y | ? | RP | MC | 30 | FF | Y | Y | Y | Y | N | 19 | $323 |
| 10 X 50 | WP-ADJUST-FREE | 235500 | 5 | Y | ? | RP | MC | 60 | FF | Y | Y | Y | Y | N | 27 | 369 |
| 7 X 35 | WATERPROOF | 215200 | 7.3 | Y | ? | RP | MC | ? | CF | Y | Y | Y | Y | N | 21 | 361 |
| 10 X 50 | WATERPROOF | 225200 | 5 | Y | ? | RP | MC | ? | CF | Y | Y | Y | Y | N | 29 | 407 |
| 8 X 24 | COMPACT | 241000 | 8.5 | Y | ? | RP | MC | ? | CF | N | Y | Y | Y | N | 10 | 229 |
| 6-15 X 24 | ZOOM | 250000 | 6.3⊕6X | Y | ? | RP | MC | ? | CF | N | Y | Y | Y | N | 16 | 417 |

125

# SELSI

| SIZE | SERIES | MODEL | FOV | RA | ER | PR | CO | NF | TOF | WP | CA | S | CPS | TA | WT | SL |
|------|--------|-------|-----|----|----|----|----|----|-----|----|----|----|-----|----|----|----|
| 10 X 20 | MINIATURES | 6 | 5 | N | ? | PP-? | C | ? | CF | N | Y | ? | ? | ? | 9 | $150 |
| 8 X 20 | MINIATURES | 7 | 5.8 | N | ? | PP-? | C | ? | CF | N | Y | ? | ? | ? | 9 | 144 |
| 8 X 21 | RP-MINI | 10 | 7 | Y | ? | RP | FC | ? | CF | Y | Y | ? | ? | ? | ? | 314 |
| 10 X 25 | RP-MINI | 18 | 5.5 | Y | ? | RP | FC | ? | CF | N | Y | ? | ? | ? | ? | 124 |
| 8 X 21 | FIXED FOCUS | 35 | 7 | Y | ? | RP | ? | ? | FF | N | ? | ? | ? | ? | ? | 150 |
| 10 X 25 | FIXED FOCUS | 36 | 5.5 | Y | ? | RP | ? | ? | FF | N | ? | ? | ? | ? | ? | 160 |
| 7 X 35 | AMERICAN SHAPE | 104 | 11 | N | ? | PP-? | FC | ? | CF | N | ? | ? | ? | ? | 26 | 150 |
| 8 X 40 | AMERICAN SHAPE | 106 | 9.8 | N | ? | PP-? | FC | ? | CF | N | ? | ? | ? | ? | 27 | 158 |
| 7 X 50 | AMERICAN SHAPE | 108 | 8 | N | ? | PP-? | FC | ? | CF | N | ? | ? | ? | ? | 33 | 162 |
| 10 X 50 | AMERICAN SHAPE | 109 | 7 | N | ? | PP-? | FC | ? | CF | N | ? | ? | ? | ? | 33 | 165 |
| 20 X 60 | HIGH POWER | 1134 | 3 | N | ? | PP-? | ? | ? | CF | N | Y | ? | ? | ? | ? | 276 |
| 20 X 70 | HIGH POWER | 1137 | 3 | N | ? | PP-? | ? | ? | CF | N | Y | ? | ? | ? | ? | 332 |
| 30 X 70 | HIGH POWER | 1143 | 2.1 | N | ? | PP-? | ? | ? | CF | N | Y | ? | ? | ? | ? | 345 |
| 8-24 X 50 | ZOOM | 1139 | 4.7@8X | N | ? | PP-? | FC | ? | CF | N | Y | Y | ? | ? | ? | 269 |
| 7-15 X 35 | ZOOM | 1140 | 5.7@7X | N | ? | PP-? | FC | ? | CF | N | Y | Y | ? | ? | ? | 193 |
| 10-30 X 60 | ZOOM | 1141 | 3.6@10X | N | ? | PP-? | FC | ? | CF | N | Y | Y | ? | ? | ? | 374 |

# SIMMONS

| SIZE | SERIES | MODEL | FOV | RA | ER | PR | CO | NF | TOF | WP | CA | S | CPS | TA | WT | SL |
|------|--------|-------|-----|----|----|----|----|----|----|----|----|----|-----|----|----|-----|
| 7-15 X 35 | RED LINE | 1109 | 5.7@7X | N | ? | PP-? | FC | ? | CF | N | Y | Y | ? | ? | ? | $145 |
| 8 X 40 | RUBBER COVERED | 1120 | 8.2 | Y | ? | PP-? | MC | ? | CF | N | Y | Y | ? | ? | ? | 101 |
| 10 X 50 | RUBBER COVERED | 1121 | 7 | Y | ? | PP-? | MC | ? | CF | N | Y | Y | ? | ? | ? | 113 |
| 7 X 50 | RUBBER COVERED | 1173 | 7.1 | Y | ? | PP-? | MC | ? | CF | N | Y | Y | ? | ? | ? | 147 |
| 8 X 25 | COMPACT | 1140 | 8.7 | Y | ? | RP | FC | ? | CF | N | Y | Y | ? | ? | ? | 112 |
| 8 X 25 | COMPACT | 1157 | 9.4 | Y | ? | RP | FC | ? | CF | N | Y | Y | ? | ? | ? | 112 |
| 10 X 21 | COMPACT | 1145 | 5.6 | Y | ? | RP | FC | ? | CF | N | Y | Y | ? | ? | ? | 110 |
| 10 X 25 | COMPACT | 1158 | 7 | Y | ? | RP | FC | ? | CF | N | Y | Y | ? | ? | ? | 122 |
| 7 X 42 | GOLD SERIES | 24180 | 6.7 | Y | ? | RP | FMC | ? | CF | N | ? | ? | ? | ? | ? | 430 |
| 10 X 42 | GOLD SERIES | 24181 | 6.5 | Y | ? | RP | FMC | ? | CF | N | ? | ? | ? | ? | ? | 452 |
| 8 X 25 | WHITETAIL CLASSIC | WCB20 | 8.9 | Y | ? | RP | MC | ? | CF | N | ? | ? | ? | ? | ? | 142 |
| 10 X 25 | WHITETAIL CLASSIC | WCB21 | 6.5 | Y | ? | RP | MC | ? | CF | N | ? | ? | ? | ? | ? | 161 |

127

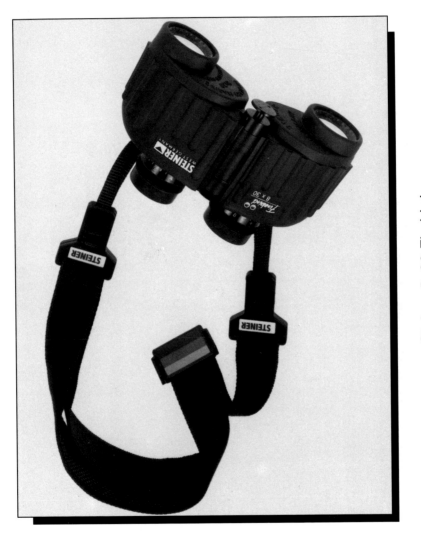

Steiner 8x30 Firebird
Photo courtesy of Pioneer Marketing & Research, Inc.

| SIZE | SERIES | MODEL | FOV | RA | ER | PR | CO | NF | TOF | WP | CA | S | CPS | TA | WT | SL |
|------|--------|-------|-----|----|----|----|----|----|-----|----|----|---|-----|----|----|----|
| 8 X 25 | CHAMP | 200 | 6.9 | N | 9 | PP-7 | MC | ? | CF | N | N | Y | N | N | 12 | $279 |
| 6 X 30 | MILITARY | 260 | 8.6 | Y | 20.2 | PP-7 | MC | ? | IF | N | N | Y | Y | N | 19 | 389 |
| 8 X 30 | MILITARY | 280 | 7.4 | Y | 11.2 | PP-7 | MC | ? | IF | N | N | Y | Y | N | 17 | 349 |
| 7 X 35 | MILITARY | 273 | 10 | Y | 13 | PP-7 | MC | ? | IF | N | N | Y | Y | N | 21 | 449 |
| 7 X 50 | MILITARY | 275 | 7 | Y | 21 | PP-7 | MC | ? | IF | N | N | Y | Y | N | 37 | 749 |
| 7 X 50 | ADMIRAL | 279 | 7 | Y | 21 | PP-4 | FMC | ? | IF | Y | N | Y | Y | N | 37 | 1049 |
| 9 X 40 | BIG HORN | 290 | 5.7 | Y | 20.2 | PP-4 | FMC | ? | IF | N | N | Y | Y | N | 26 | 649 |
| 10 X 50 | MILITARY | 210 | 6.2 | Y | 16.8 | PP-7 | MC | ? | IF | N | N | Y | N | N | 36 | 749 |
| 8 X 56 | NIGHT HUNTER | 485 | 5.7 | Y | 15.5 | PP-4 | FMC | ? | IF | Y | Y | Y | Y | N | 42 | 1049 |
| 15 X 80 | OBSERVER | 415 | 3.7 | Y | 10.7 | PP-4 | FMC | ? | IF | Y | N | Y | Y | Y | 56 | 1449 |
| 8 X 30 | FIREBIRD | 183 | 7.4 | Y | 11.2 | PP-7 | MC | ? | IF | N | N | Y | Y | N | 17 | 399 |
| 7 X 50 | POLAR BEAR | 274 | 7.3 | Y | 21 | PP-4 | FMC | ? | IF | Y | N | Y | Y | N | 44 | 999 |
| 6 X 30 | WHITETAIL | 402 | 8.6 | Y | 20.2 | PP-4 | FMC | ? | IF | Y | N | Y | Y | N | 19 | 799 |
| 7 x 50 | WHITETAIL | 405 | 7 | Y | 21 | PP-4 | FMC | ? | IF | Y | N | Y | Y | N | 37 | 1149 |

Swarovski 8x56 SL
Photo courtesy of Swarovski Optik

# SWAROVSKI

| SIZE | SERIES | MODEL | FOV | RA | ER | PR | CO | NF | TOF | WP | CA | S | CPS | TA | WT | SL |
|---|---|---|---|---|---|---|---|---|---|---|---|---|---|---|---|---|
| 6 X 30 | TRADITIONAL | B 6 X 30-M | 8.5 | N | 13 | PP-4 | FMC | 15 | CF | Y | Y | Y | N | N | 18 | $445 |
| 8 X 30 | TRADITIONAL | B 8 X 30 N-M | 7 | N | 12 | PP-4 | FMC | 18 | CF | Y | Y | Y | N | N | 18 | 495 |
| 8 X 30 | TRADITIONAL | B 8 X 30 N-MGA | 7 | Y | 12 | PP-4 | FMC | 18 | CF | Y | N | Y | Y | N | 20 | 625 |
| 8 X 30 | TRADITIONAL | B 8 X 30 W-M | 7.9 | N | 12 | PP-4 | FMC | 18 | CF | Y | Y | Y | N | N | 19 | 515 |
| 8 X 30 | TRADITIONAL | B 8 X 30 W-MGA | 7.9 | Y | 12 | PP-4 | FMC | 18 | CF | Y | N | Y | Y | N | 20 | 650 |
| 7 X 42 | TRADITIONAL | B 7 X 42 B-OGA | 6.5 | Y | 14 | PP-4 | FMC | 16 | IF | Y | N | Y | Y | N | 26 | 715 |
| 7 X 42 | TRADITIONAL | B 7 X 42-M | 6.5 | N | 14 | PP-4 | FMC | 16 | CF | Y | Y | Y | N | N | 23 | 490 |
| 7 X 42 | TRADITIONAL | B 7 X 42-MGA | 6.5 | Y | 14 | PP-4 | FMC | 16 | CF | Y | N | Y | Y | N | 26 | 640 |
| 10 X 40 | TRADITIONAL | B 10 X 40-M | 6.3 | N | 12 | PP-4 | FMC | 13 | CF | Y | Y | Y | N | N | 24 | 565 |
| 10 X 40 | TRADITIONAL | B 10 X 40-MGA | 6.3 | Y | 12 | PP-4 | FMC | 13 | CF | Y | N | Y | Y | N | 27 | 715 |
| 7 X 42 | SL | B 7 X 42-SL-BA | 6.5 | N | 13 | PP-4 | FMC | 16 | CF | Y | N | Y | Y | N | 31 | 740 |
| 7 X 50 | SL | B 7 X 50-SL-Y | 7.1 | N | 21.5 | PP-4 | FMC | 19 | CF | Y | N | Y | Y | N | 38 | 840 |
| 8 X 56 | SL | B 8 X 56-SL-BA | 5.8 | N | 18 | PP-4 | FMC | 20 | CF | Y | N | Y | Y | N | 43 | 950 |
| 10 X 40 | SL | B 10 X 40W-SL-BA | 6.3 | N | 13.5 | PP-4 | FMC | 13 | CF | Y | N | Y | Y | N | 31 | 830 |
| 10 X 50 | SL | B 10 X 50-SL-BA | 5.8 | N | 13.5 | PP-4 | FMC | 20 | CF | Y | N | Y | Y | N | 36 | 915 |
| 7 X 30 | SLC | B 7 X 30B-SL-C-BA | 7.4 | N | 18 | RP | FMC | 15 | CF | Y | Y | Y | Y | N | 19 | 600 |
| 8 X 30 | SLC | B 8 X30W-SL-C-BA | 7.8 | N | 15.5 | RP | FMC | 15 | CF | Y | Y | Y | Y | N | 19 | 610 |
| 8 X 20 | ? | B 8 X 20 B-P | 6.6 | N | 13 | RP | FMC | 13 | CF | Y | Y | Y | ? | N | 8 | 470 |

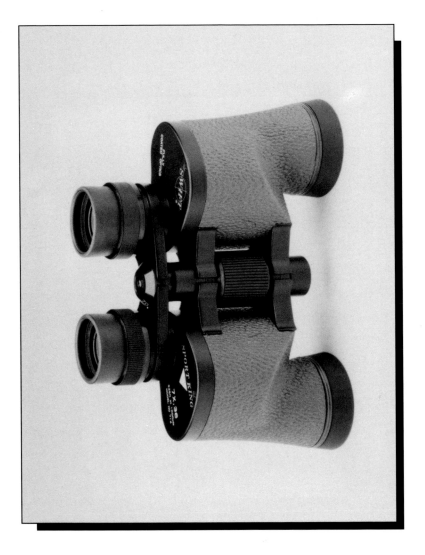

Swift 7x35 Sport King
Photo courtesy of Swift Instruments Inc.

# SWIFT

| SIZE | SERIES | MODEL | FOV | RA | ER | PR | CO | NF | TOF | WP | CA | S | CPS | TA | WT | SL |
|------|--------|-------|-----|----|----|----|----|----|-----|----|----|----|-----|----|----|-----|
| 10 X 50 | COUGAR | 704 | 7 | Y | 13 | PP-7 | FC | 20 | CF | N | Y | Y | Y | Y | 29 | $173 |
| 8 X 40 | FALCON | 702 | 9 | Y | 14 | PP-7 | FC | 15 | CF | N | Y | Y | Y | Y | 25 | 160 |
| 7 X 35 | ALL-SPORT | 701 | 11 | Y | 12 | PP-7 | FC | 12 | CF | N | Y | Y | Y | Y | 23 | 150 |
| 7 X 50 | DOLPHIN | 703 | 7.1 | Y | 18 | PP-7 | FC | 18 | CF | N | Y | Y | Y | Y | 29 | 170 |
| 7 X 35 | C-AUDUBON | 825 | 8.2 | Y | 11.5 | RP | MC | 13 | CF | Y | Y | Y | Y | N | 21 | 575 |
| 8.5 X 44 | AUDUBON | 804 | 8.2 | N | 15 | PP-4 | MC | 13 | CF | N | Y | Y | Y | Y | 29 | 375 |
| 10 X 42 | CONDOR | 719 | 7 | N | 10.5 | PP-4 | FC | 15 | CF | N | Y | Y | Y | Y | 28 | 300 |
| 8 X 40 | TRILYTE | 806 | 6.5 | Y | 16.5 | RP | MC | 24 | CF | N | N | Y | Y | N | 22 | 280 |
| 8 X 40 | HERON | 739 | 9 | N | 11 | PP-7 | FC | 18 | CF | N | Y | Y | Y | Y | 25 | 130 |
| 7.5 X 42 | OSPREY | 754 | 7 | Y | 23 | PP-4 | FC | 14 | CF | N | N | Y | Y | Y | 30 | 315 |
| 7 X 35 | NEPTUNE | 802 | 8 | N | 15 | PP-4 | FC | 15 | CF | N | Y | Y | Y | Y | 21 | 255 |
| 7 X 50 | SEA WOLF | 713 | 7.1 | Y | 19 | PP-7 | FC | 24 | CF | Y | N | Y | Y | Y | 33 | 253 |
| 7.5 X 44 | MARLIN | 715 | 7.3 | Y | 16 | PP-4 | FC | 15 | IF | Y | N | Y | Y | Y | 39 | 450 |
| 7 X 50 | SEA HAWK | 753 | 7.5 | Y | 19 | PP-4 | FC | 30 | IF | Y | Y | Y | Y | Y | 36 | 375 |
| 7 X 50 | SKIPPER | 789 | 7.3 | N | 16 | PP-4 | FC | 18 | CF | Y | N | Y | Y | Y | 31 | 240 |
| 7 X 50 | STORM KING | 717 | 7.3 | N | 16 | PP-4 | FC | 21 | IF | Y | Y | Y | Y | Y | 52 | 600 |
| 7 X 50 | ASF | 781R | 8 | Y | 17 | PP-7 | FC | 13 | CF | N | Y | Y | Y | N | 32 | 138 |
| 10 X 25 | TRILYTE | 824 | 7 | Y | 4.8 | RP | FC | 14 | CF | N | Y | Y | N | N | 12 | 135 |
| 7 X 21 | TRILYTE | 800 | 9.3 | N | 8 | RP | FC | 13 | CF | N | Y | Y | Y | N | 8 | 125 |
| 8 X 25 | TRILYTE | 821 | 8.7 | Y | 5.3 | RP | FC | 10 | CF | N | Y | Y | N | N | 10 | 130 |
| 7 X 35 | SPORT KING | 714 | 10 | N | 13.5 | PP-4 | FC | 12 | CF | N | Y | Y | Y | Y | 25 | 290 |
| 7-15 X 35 | SF ZOOM | 774 | 5.7@7X | N | 13/11 | PP-7 | FC | 18 | CF | N | Y | Y | Y | Y | 23 | 205 |
| 8 X 40 | NIGHT HAWK | 771 | 9.5 | N | 12 | PP-7 | FC | 18 | CF | N | Y | Y | N | Y | 28 | 185 |
| 11 X 80 | OBSERVER | 845 | 4.5 | N | 15 | PP-4 | FC | 60 | CF | N | Y | Y | Y | N | 78 | 550 |
| 20 X 80 | SATELLITE | 846 | 3.5 | N | 14.5 | PP-4 | FC | 60 | CF | N | Y | Y | Y | Y | 80 | 560 |
| 10 X 25 | MICRON | 801 | 5 | Y | ? | RP | MC | 15 | CF | N | Y | Y | N | N | 9 | 124 |
| 6-12X 23 | MICRON | 803 | ? | Y | ? | RP | FC | ? | CF | N | Y | Y | N | N | 12 | 200 |

# TASCO

| SIZE | SERIES | MODEL | FOV | RA | ER | PR | CO | NF | TOF | WP | CA | S | CPS | TA | WT | SL |
|---|---|---|---|---|---|---|---|---|---|---|---|---|---|---|---|---|
| 9 X 25 | INFOCUS | 89925FI | 5 | ? | ? | RP | FC | 66 | FF | N | Y | Y | ? | ? | 9 | $120 |
| 7 X 21 | INFOCUS | 89721FI | 6.5 | ? | ? | RP | FC | 40 | FF | N | Y | Y | ? | ? | 7 | 112 |
| 7 X 35 | INFOCUS | 89735WAFI | 9.5 | ? | ? | PP-? | FC | 40 | FF | N | Y | Y | ? | ? | 24 | 128 |
| 6-21 X 35 | INFOCUS | 8961235FZ | 6.5@6X | ? | ? | PP-? | FC | 29 | FF | N | Y | Y | ? | ? | 24 | 192 |
| 7 X 50 | INFOCUS | 322BWFI | 7 | ? | ? | PP-? | FC | 40 | FF | Y | Y | Y | ? | Y | 35 | 256 |
| 7 X 50 | INFOCUS | 89750 | 7.1 | ? | ? | PP-? | FC | 40 | FF | N | Y | Y | ? | ? | 28 | 144 |
| 8 X 22 | INFOCUS | 89822BC | 7.1 | ? | ? | RP | FC | 52 | FF | N | Y | Y | ? | ? | 9 | 112 |
| 8 X 25 | INFOCUS | 89825WA | 8.9 | ? | ? | RP | FC | 52 | FF | N | Y | Y | ? | ? | 9 | 144 |
| 10 X 26 | INFOCUS | 391026BC | 5.4 | ? | ? | RP | FC | 81 | FF | N | Y | Y | ? | ? | 9 | 120 |
| 10 X 50 | INFOCUS | 391050WA | 6.5 | ? | ? | PP-? | FC | 81 | FF | N | Y | Y | ? | ? | 28 | 148 |
| 7-14 X 40 | INFOCUS | 3971440FZ | 5.6@7X | ? | ? | PP-? | FC | 40 | FF | N | Y | Y | ? | Y | 24 | 208 |
| 8-16 X 50 | INFOCUS | 3981650FZ | 4.7@8X | ? | ? | PP-? | FC | 52 | FF | N | Y | Y | ? | Y | 28 | 224 |
| 8-24 X 50 | WORLD CLASS | 107BRZ | 5@8X | Y | ? | PP-4 | MC | ? | CF | N | Y | Y | Y | Y | 36 | 240 |
| 8 X 25 | WORLD CLASS | 167RB | 8.9 | Y | ? | RP | MC | ? | CF | N | Y | Y | Y | ? | 10 | 136 |
| 7-15 X 35 | ZOOM | 101ZC | 5.5@7X | N | ? | PP-? | C | ? | CF | N | Y | Y | ? | ? | 21 | 128 |
| 7-21 X 40 | ZOOM | 102Z | 5.5@7X | N | ? | PP-? | FC | ? | CF | N | Y | Y | ? | Y | 25 | 168 |
| 7-21 X 40 | ZOOM | 102BRZ | 5.5@7X | Y | ? | PP-? | FC | ? | CF | N | Y | Y | ? | Y | 29 | 184 |
| 8-20 X 50 | ZOOM | 108BRZ | 4.7@8X | Y | ? | PP-? | FC | ? | CF | N | Y | Y | ? | Y | 33 | 200 |
| 7-35 X 50 | ZOOM | 117BRZ | 4.1@7X | Y | ? | PP-? | FC | ? | CF | N | Y | Y | ? | Y | 37 | 288 |
| 8 X 25 | WATERPRCOF | 171BRW | 8.2 | Y | ? | RP | FC | ? | CF | Y | Y | Y | ? | Y | 13 | 256 |
| 7 X 50 | WATERPRCOF | 322BCW | 7 | Y | ? | PP-4 | FC | ? | IF | Y | Y | Y | ? | Y | 37 | 344 |
| 8 X 30 | WATERPRCOF | 327MW | 8.5 | Y | ? | PP-4 | FC | ? | IF | Y | Y | Y | ? | Y | 27 | 680 |
| 7 X 50 | WATERPRCOF | 328MW | 7.3 | Y | ? | PP-4 | FC | ? | IF | Y | Y | Y | ? | Y | 52 | 760 |
| 7 X 42 | COMPACT | 151BLE | 7.1 | Y | ? | RP | FC | ? | CF | N | Y | Y | ? | ? | 24 | 240 |
| 7-15 X 25 | COMPACT | 200RB | 5.3@7X | Y | ? | PP-? | FC | ? | CF | N | Y | Y | ? | ? | 11 | 160 |

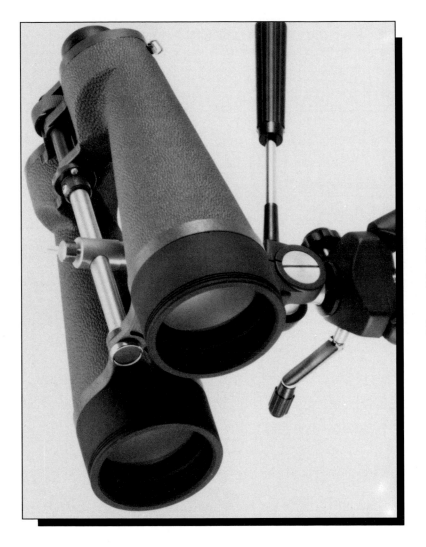

Vixen 30x80 BCF

Photo courtesy of Vixen Optical Industries Ltd.

# VIXEN

| SIZE | SERIES | MODEL | FOV | RA | ER | PR | CO | NF | TOF | WP | CA | S | CPS | TA | WT | SL |
|---|---|---|---|---|---|---|---|---|---|---|---|---|---|---|---|---|
| 8 x 23 | COMPACT | 1312 | 6.3 | N | 15 | PP-4 | FC | 16 | CF | N | Y | Y | Y | N | 12 | $89 |
| 10 X 24 | COMPACT | 1322 | 5 | N | 11 | PP-4 | FC | 23 | CF | N | Y | Y | N | N | 12 | 99 |
| 8 X 24 | COMPACT | 1308 | 7.5 | N | 10 | PP-4 | FC | 16 | CF | N | Y | Y | N | N | 9 | 79 |
| 10 X 24 | COMPACT | 1309 | 6.5 | N | 8.5 | PP-4 | FC | 23 | CF | N | Y | Y | N | N | 9 | 79 |
| 6-12 X 25 | COMPACT | 1531 | 5.3@7X | N | 16.5 | PP-7 | MC | 18 | CF | N | Y | Y | N | N | 17 | 99 |
| 8 X 20 | DCF | 1602 | 7.2 | N | 6.5 | RP | C | 23 | CF | N | Y | Y | Y | N | 6 | 109 |
| 9 X 20 | DCF | 1603 | 6.4 | N | 6 | RP | C | 30 | CF | N | Y | Y | Y | N | 6 | 109 |
| 8 X 30 | DCF | 1604 | 7 | N | 12.5 | RP | C | 26 | CF | N | Y | Y | Y | N | 14 | 135 |
| 10 X 30 | DCF | 1605 | 5.6 | N | 9.5 | RP | C | 36 | CF | N | Y | Y | Y | N | 18 | 139 |
| 7.5 X 42 | BIF | 1410 | 7 | Y | 24 | PP-7 | FC | 23 | IF | N | N | Y | Y | Y | 29 | 139 |
| 10 X 70 | BCF | 1403 | 4.8 | N | 18 | PP-4 | MC | 82 | CF | N | Y | Y | Y | Y | 65 | 299 |
| 30 X 80 | BCF | 1409 | 2.1 | N | 6 | PP-4 | MC | 59 | CF | N | Y | Y | Y | Y | 80 | 369 |
| 20 X 100 | BCF | 1413 | 2.5 | N | 16.4 | PP-4 | MC | 89 | CF | N | Y | Y | Y | Y | 116 | 899 |
| 7 X 50 | WATERPROOF | 1411 | 7.3 | N | 17 | PP-4 | FC | 39 | IF | Y | Y | Y | Y | Y | 46 | 269 |
| 10 X 70 | WATERPROOF | 1412 | 5.3 | N | 15 | PP-4 | FC | 82 | IF | Y | Y | N | Y | Y | 88 | 499 |
| 33-50 X 60 | PASTORAL | 9829 | 1.4 | N | 15 | PP-7 | MC | 32 | IF | N | Y | N | Y | Y | 123 | 999 |

# ZEISS

| SIZE | SERIES | MODEL | FOV | RA | ER | PR | CO | NF | TOF | WP | CA | S | CPS | TA | WT | SL |
|---|---|---|---|---|---|---|---|---|---|---|---|---|---|---|---|---|
| 6 X 20 | MINI | 522006 | 6.6 | ? | ? | RP | FC | ? | IF | N | Y | ? | ? | ? | 5 | $430 |
| 8 X 20 | MINI | 522024 | 6.6 | ? | ? | RP | FC | ? | CF | N | Y | ? | ? | ? | 6 | 480 |
| 10 X 25 | MINI | 522025 | 5.4 | ? | ? | RP | FC | ? | CF | N | Y | ? | ? | ? | 7 | 525 |
| 8 X 30 | STANDARD | 523508 | 7.7 | N | ? | RP | FMC | ? | CF | N | Y | ? | ? | ? | 19 | 970 |
| 10 X 40 | STANDARD | 524012 | 6.3 | N | ? | RP | FMC | ? | CF | N | Y | ? | ? | ? | 24 | 1190 |
| 8 X 20 | RUBBER | 522030 | 6.6 | Y | ? | RP | FC | ? | CF | N | Y | ? | ? | ? | 9 | 580 |
| 7 X 42 | RUBBER | 524003 | 8.6 | Y | ? | RP | FMC | ? | CF | N | Y | ? | ? | ? | 28 | 1130 |
| 7 X 50 | RUBBER | 525505 | 7.4 | Y | ? | PP-? | FMC | ? | IF | Y | Y | ? | ? | ? | 42 | 1331 |
| 8 X 30 | RUBBER | 523514 | 6.6 | Y | ? | RP | FC | ? | CF | N | N | ? | ? | ? | 21 | 845 |
| 8 X 30 | RUBBER | 523509 | 7.7 | Y | ? | RP | FMC | ? | CF | N | Y | ? | ? | ? | 20 | 970 |
| 8 X 56 | RUBBER | 525656 | 6.3 | Y | ? | RP | FMC | ? | CF | N | Y | ? | ? | ? | 36 | 1380 |
| 10 X 40 | RUBBEF | 524013 | 6.3 | Y | ? | RP | FMC | ? | CF | N | Y | ? | ? | ? | 26 | 1190 |
| 15 X 60 | RUBBEF | 520602 | 4.6 | Y | ? | PP-? | FMC | ? | CF | N | N | ? | ? | ? | 56 | 2005 |

139

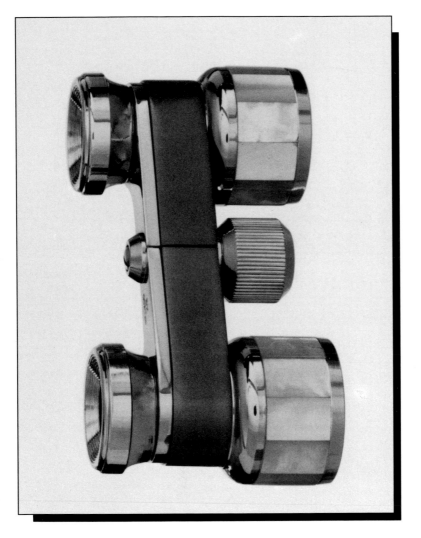

Bushnell 3x23 Theatre Glass
Photo courtesy of Bushnell

# OPERA GLASSES

| SIZE | MFG | SERIES | MODEL | OO | CA | S | WT | SL |
|------|-----|--------|-------|----|----|---|----|----|
| 3.6 X 12 | ZEISS | THEATERGLASS | 520470 | FC | Y | Y | 6 | $720 |
| 2.5 X 16 | AUS JENA | LUXOSTAR | OPERA | FC | Y | N | 3 | 100 |
| 3 X 23 | BUSHNELL | CONCERT | 15-3230 | C | Y | Y | 12 | 137 |
| 3 X 25 | CELESTRON | OPERA GLASS | 71500 | FC | Y | Y | 5 | 75 |
| 3 X 25 | PARKS | ? | 401-1000 | FC | Y | N | ? | 150 |
| 3 X 25 | SELSI | OPERA | 213 | ? | Y | ? | ? | 177 |
| 3 X 25 | SWIFT | BARONESS | 706 | ? | Y | ? | ? | 162 |
| 3 X 25 | SWIFT | DUCHESS | 708 | ? | Y | ? | ? | 150 |
| 3 X 25 | SWIFT | PRINCESS | 813 | ? | Y | ? | ? | 75 |
| 3 X 25 | TASCO | ? | 595 | ? | ? | ? | 5 | 128 |
| 5 X 25 | BUSHNELL | CONCERT | 15-0525 | C | Y | Y | 10 | 91 |
| 3 X 27 | TASCO | ? | 589 | ? | ? | ? | 7 | 88 |
| 3 X 28 | BUSHNELL | CONCERT | 15-3287 | C | Y | Y | 12 | 137 |
| 4 X 30 | SWIFT | LARK | 811 | C | Y | ? | ? | 140 |
| 4 X 30 | VIXEN | OPERA GLASS | 1206 | C | Y | N | 7 | 55 |

141

Celestron 8x25 Monocular
Photo courtesy of Celestron International

# MONOCULARS

| SIZE | MFG | SERIES | MODEL | FOV | RA | PR | OO | CA | S | CPS | WT | SL |
|------|-----|--------|-------|-----|----|----|----|----|---|-----|----|-----|
| 5 X 20 | CELESTRON | GOLFING MONOCULAR | 71502 | 7.3 | N | RP | FC | Y | Y | Y | 2.3 | $110 |
| 5 X 20 | TASCO | GOLF SCOPE | 514 | 7.3 | N | RP | ? | ? | ? | ? | 3.5 | 88 |
| 6 X 20 | ZEISS | MINI | 522009 | 6.9 | N | RP | FC | Y | ? | ? | 1.7 | 225 |
| 7 X 20 | PENTAX | MONOCULAR | 61305 | 7.5 | N | RP | MC | Y | Y | Y | 2 | 120 |
| 8 X 20 | BUSHNELL | SPORTVIEW | 14-8200 | 6.5 | N | RP | FC | Y | Y | N | 2.1 | 110 |
| 8 X 20 | SELSI | ? | 163 | ? | N | PP-? | C | Y | ? | ? | ? | 74 |
| 8 X 20 | SELSI | ? | 164 | 7 | N | RP | C | Y | Y | ? | ? | 134 |
| 8 X 20 | TASCO | ? | 516 | 5.5 | N | PP-? | FC | ? | Y | ? | 2.1 | 100 |
| 8 X 20 | VIXEN | MONOCULAR | 1102 | 7.2 | N | RP | FC | Y | Y | Y | 2.5 | 55 |
| 8 X 20 | ZEISS | MINI | 522005 | 6.9 | N | RP | FC | Y | ? | ? | 1.7 | 225 |
| 10 X 20 | SELSI | ? | 160 | ? | N | PP-? | C | Y | ? | ? | ? | 76 |
| 8 X 21 | AUS JENA | ? | TURMON | 7 | N | PP-4 | MC | Y | N | N | 5 | 210 |
| 8 X 21 | SIMMONS | COMPACT | 1164 | ? | ? | RP | FC | Y | ? | ? | ? | 57 |
| 8 X 21 | TASCO | COMPACT | 565RB | 7.3 | Y | RP | FC | ? | Y | ? | 2.8 | 48 |
| 10 X 21 | SIMMONS | COMPACT | 1160 | 7 | Y | RP | FC | Y | ? | ? | ? | 62 |
| 8 X 22 | OPTOLYTH | SPORTING | 15020 | 6.7 | N | RP | FMC | Y | N | Y | 2.5 | 225 |
| 8 X 24 | BUSHNELL | SPORTVIEW | 14-6180 | 8 | N | PP-7 | FC | Y | Y | N | 4.4 | 115 |
| 7 X 25 | SELSI | ? | 168 | 10 | N | RP | C | Y | Y | ? | ? | 200 |
| 8 X 25 | CELESTRON | MINI | 71504 | 8.7 | Y | RP | MC | Y | Y | Y | 4 | 60 |
| 8 X 25 | PARKS | ? | 401-10020 | 8.7 | Y | RP | MC | Y | Y | Y | ? | 100 |
| 8 X 25 | SWIFT | ARMORED | 778 | 8.7 | Y | RP | ? | Y | Y | N | ? | 85 |
| 6 X 30 | SELSI | ? | 150 | 7.5 | N | PP-? | ? | Y | ? | ? | ? | 74 |
| 8 X 30 | PENTAX | MONOCULAR | 62202 | 6.2 | N | RP | MC | Y | Y | Y | 6 | 200 |
| 8 X 30 | SWIFT | TRILYTE | 777 | 8.5 | N | RP | C | Y | Y | Y | 6.4 | 200 |
| 8 X 30 | VIXEN | MONOCULAR | 1103 | 7 | N | RP | FC | Y | Y | Y | 5.3 | 75 |
| 10 X 30 | BUSHNELL | SPORTVIEW | 14-1031 | 6.7 | Y | RP | FC | Y | N | N | 8 | 163 |
| 10 X 30 | SELSI | ? | 166 | 6 | N | RP | C | Y | Y | ? | ? | 176 |

# CHAPTER 19

## SUMMARY

I hope you have expanded your knowledge of binoculars after reading this book.

Any comments from you would be appreciated.  Especially helpful would be any corrections or ideas to expand coverage in areas not touched upon.

Please address comments to:

Alan Hale
P.O. Box 3578
Torrance, CA  90510